JACKETING BOTTLES AND
OTHER CONTAINERS

PLATE 1. Calabash Punch Bowl with Oriental Weave Jacket for a Lawn Party.

VIVA J. COOKE

JACKETING BOTTLES

AND OTHER CONTAINERS

E. A. SEEMANN PUBLISHING, INC.

Miami, Florida

To the memory of

a Father who demanded the learning of new
      skills and
a Mother who encouraged manual
      experiment and practice,

this effort is gratefully dedicated.

CONTENTS

ILLUSTRATIONS

FOREWORD

THE INSTRUCTIONS in this book may seem overly detailed, but they were written with the inexpert worker in mind, who generally does better if no step in directions is left to the imagination. This is particularly true of one working without other help than the printed page. Most of the instructions are given in the form of a step to a paragraph so as to make them easier to follow.

Forms, materials, and procedures are so varied that the instructions given for these examples are meant to be suggestive as well as specific. Nearly all of them may be adapted to other forms, used with other mediums, or in combinations of different materials.

The forms shown probably cannot be duplicated exactly, but given the height and diameter of a form, the width and length of spokes and weavers used to make the jacket, with suggestions for suitable materials, the worker should be able to adapt material and procedure to the lines of an appropriate form.

A straight-line shape and a simple procedure such as the over-one, under-one or check weave should be chosen for an initial project. This will give experience in preparing and handling materials, familiarize the worker with basic terms, and give practice essential to a growth of confidence and skill.

These directions are the result of a determination formed years ago when I was unable to find anything comprehensive on bottle jackets. I hope they will be as helpful to others as I intend them to be and that they will give as much pleasure and satisfaction as I have found in bringing the idea to fruition.

Viva Jane Cooke

SOME HELPFUL TOOLS

THE JACKETING of bottles, like basketry, is largely a manual skill in which soft materials are used, seldom calling for heavy tools. The tools needed are simple and, except for some of the needles, will be found in most households, or at least be readily obtainable. The peg holder is made of a one-foot and a two-foot length of 2 x 4, and the dowels of scraps. A section of old broom handle is just right for the largest peg when sawed to about 8 inches.

To make the peg holder, saw two cuts at the center of each 2 x 4 a distance of 4 inches apart and exactly 1 inch deep.

Remove the wood between the cuts evenly to make a right-angle mortise, allowing the sections to lie flush when one is fitted into the other. Hold with a screw at the center.

Bore holes of a size to fit the various dowels about three fourths of the way through the holder, making the largest hole near the center cut to provide a firm base for a large inverted bottle. If the dowels need adjusting to the holes, they may be reduced at the base for the depth of the hole. They should fit firmly but be removable when not in use.

The really indispensable tool is a bag needle, with its flattened, curved end which will serve various purposes. The curved mattress needle will be invaluable in inserting and bringing out strands around necks and other closely woven flat surfaces. The raffia needle has its place in pine-needle work, and the leather needle with its tongue will be helpful with flat strands. These are recommended as needed in the various jacket procedures.

A pair of pliers is sometimes useful to pull the needle which carries the binder through and spring clothespins.

PLATE 2. Some Useful Tools.

CHAPTER TWO

THE OVER-ONE, UNDER-ONE OR CHECK-WEAVE PROCEDURE

This procedure is the simplest form of weave and comes first because it is helpful in teaching one to handle the jacketing materials, to manipulate them in relation to a form, to obtain a tight weave, and other facets of the work which only experience will give. Even if the worker is not interested in the pattern, the time will still be profitably used, for a more intricate pattern will later be easier to follow, and the jacket more skillfully made because of the practice.

The check weave is almost invariably made with a continuous strand; that is, the weaver or horizontal strand which fastens the spokes or vertical strands together passes in rows around the bottle on a slightly ascending angle. To make this possible, there must be an uneven number of spoke strands so that the weaver will automatically alternate in weaving the spokes, thus giving them an over-one, under-one weave.

In the instructions and the examples given, the spoke strands are most frequently fastened at their centers for the base weave, and are made to ray out for a jacket base and up to form the spokes for the side weave. Of course this makes an even number of spokes for any combination, so an extra spoke is either inserted in the base weave or, most often, left at the left of the weaver as a long extension which is treated as a spoke when weaver and spoke are of the same material.

The pattern may be in small square checks, in which case the weaver and the spoke are usually, though not necessarily, of the same material, and certainly of the same width. There should be a sufficient number of spokes to cover the sides of the form with little space between them, which would require a bottle with straight sides.

There are infinite variations in size of spoke and width of weaver, in combination of spoke and weaver materials, and in intervals and widths of form exposure between spaced spokes and weavers.

This last is particularly true in jackets for forms which are smaller at the base or top. Each form requires individual adaptations. This weave accommodates itself very readily to any flare and the subsequent drawing in at the shoulder and neck.

Occasionally a shaped spoke is used which follows the shape of the form to such a degree that it widens or diminishes accordingly, yet the weave is still over one, under one, resulting in a patterned unevenness which is both attractive and unusual.

Jacket of Flat Reed for a Straight Bottle

This closely woven jacket with a round reed ropelike handle turns a mere container into an object of interest and simple beauty. (Fig. 1, Plate 3).

The bottle is 16 inches tall by 5 ½ inches diameter. A specific spoke length and number are given for this particular jacket, but the weave will adapt to a rounded surface of any size and shape.

Both spokes and weavers are of 3/16-inch flat reed and as the spokes pass entirely around the bottle, a strand length of 42 inches should be provided for the nine pairs of spokes. As an uneven spoke number is required for a continuous weave, a nineteenth spoke about 23 inches long should be provided.

Coil the strands loosely and tie them for soaking about thirty minutes in cold water. They should then be wrapped in a wet towel to lie until thoroughly pliable.

Place the centers of four spoke strands side by side and hold them with a spring clothespin.

Lay the centers of three other spoke strands over these at a right angle and hold all together with clothespins.

Place the end of the weaver under a spoke and hold it with a clothespin until it can be caught in the weave which fastens the groups together by weaving over and under them as groups.

Separate one group so as to form an uneven number of spokes and put in two rows of weave around the groups, catching in the weaver end, to make a smooth, flat base grouping of fourteen spokes.

Slip the end of the extra spoke neatly under this weave and continue the weave, separating the groups into smaller groups and as soon as possible into individual spokes, using an over-one, under-one routine.

There are two strands or four spokes yet to be included in this base mat. They were left out of the original groups in order to cut down on the width and thickness at the center. Spacing for these pairs should be left on opposite sides of the base mat as the adding of two spokes at once does not disturb the sequence of the weave. Bend them at the center and insert them into the weave in the spaces reserved for them, as the expansion of the base mat allows. Work toward an equidistant spacing of spokes by the time the base mat will cover the bottom of the bottle.

With the base mat a trifle larger than the bottle base, tie mat to bottle with soft, strong string, from neck over base mat and back to neck on four sides, being sure that centers of mat and bottle are even. Use of a peg holder will make the weaving from this point to the shoulder easier, though successful weaving may be done with the bottle upright on a table.

Splice the weaver as necessary by slipping the end of a new weaver over the old one about 3 inches from its end and weaving them as one.

Pull the spokes to an evenly spaced vertical position at the edge of the bottle and fasten them firmly in place with the weave which should be tight with the weaver edges touching. Examine frequently to see that the spokes hold a true vertical and that the weave is tight and close. Any corrections in the work must be made before the shoulder is turned; a very meticulous examination is recommended when that point is reached.

The natural taper of the bottle above the shoulder will determine the space between spokes which will become less than ¼ inch at the base of the neck, though it remains for the worker to keep these spaces even.

At the fourth row of weave above the base of the neck, begin to combine the spokes and change the weave to over two, under two. Keep to this routine until the diminishing size of the neck makes it possible to place the pairs of spokes, right over left, with a spacing of ¼ inch between spokes, when the weave will again be over one, under one.

At about 1 ¼ inches below the mouth of the bot-

PLATE 3. Simple Over and Under Weaves:
Flat reed jacket with handle for large straight bottle (fig. 1); flask jacket of split soft rush, with open weave around neck (fig. 2); rectangular bottle jacket of flat reed (fig. 3); sweet grass jacket for vase woven over shaped spokes (fig. 4); raffia woven over cane spokes for a small straight bottle (fig. 5); split willow jacket for a bulbous shape (fig. 6).

tle, finish off the weave smoothly, fasten the end of the weaver, and run it down into the weave as invisibly as possible.

Before any manipulation for finishing off the spoke ends is attempted, they should be made thoroughly pliable by allowing them to lie in a wet towel.

The upper end of the handle is looped around the neck just above the upper edge of the jacket, so the finish of coiled spokes should be tight and flat so as not to interfere with the handle strands. The edge made as follows will serve this purpose.

Coil the lower spoke of a pair lightly, pass it across its own upper spoke and behind the pair on its right, where it is slipped down into the weave. Bring the end to the outside between spokes so that the small scallop so made may be evenly adjusted when all lower spokes have been woven in, by pulling them down tightly. This leaves the upper spokes of the pairs to be trimmed off close to the upper edge of the scallop.

The weaving in of the ends will be less difficult if the end is cut to a point and the weave opened up for it with a big needle, or it may be inserted by threading it in a bag needle used head first.

For the applied handle, provide a 76-inch length of No. 7 round reed. This will likely be a little long but will provide ease for finishing. Moisten it as for flat reed.

This procedure, known as a rope handle, is made over a foundation formed by itself and is accomplished by successive rows of wrapping up and down the foundation to fill in the spaces left by the natural angle of the wrapping strand.

Whittle one end of the reed to a point and run it down into the jacket weave invisibly on the right side of the selected spoke about an inch below the point of the shoulder or 7 inches below the top of the bottle. Carry the strand to the upper edge of the jacket on an easy angle to the right, around the neck, and under itself.

Wrap over and under this handle base twice firmly, but allow the strand to adjust itself to the foundation. This wrapping will set the pattern for the rope weave so should be very carefully put in.

At about an inch below the point where the end of the reed was run under the jacket weave, pass the end of the reed strand under the jacket spoke on its right side and out from under the row of weave in line with the inserted first end. This makes the first of two loops around the lower end of the handle.

Carry the weaver to the right over the handle base and twice over and under it, adjusting the wrappings to the pattern set by the first wrapping. Carry the reed strand around the neck close above the foundation strand.

Pass the strand under the handle, wrap it twice over and under, fitting the wrappings into the pattern smoothly to bring the end out below the lower end of the handle on the right.

Lay the strand smoothly in a loop following the right side of the first loop, down below the lower end of it, over to the left side of the spoke, and under the jacket weave one row of weave lower down. Bring the end out from under the weave on the left side of the first loop in line with it. Pull to adjust the loop symmetrically with the lower end centered over the spoke.

Bring the strand to the right side of the handle and wrap over and under in pattern twice, placing to fill in spaces. Carry the strand to the right, around the neck, under itself, and back to the base of the handle with wrappings to fill in the remaining spaces.

This will bring the strand to the outer left just above the two strands forming the upper left loops. Slip the end under these loop strands and across under the right loops which will hold it in place. Cut it off close to the outer edge of the right loop when dry.

Jacket in Combined Check and Open Weave

The material used for this jacket is known as soft rush, presumably from its large inner core of pith, for certainly there is nothing soft about its exterior. If a finger comes in contact with a stripped edge at just the right angle when one is preparing the blade or rod for use, it will cut as readily as a sharp knife. The plant, known in some areas as jewel rush, grows in or near water in rods as much as ¼ inch in diameter and 5-6 feet high. The bloom is unusual, growing out from the side of the rod a few inches below the tip. It splits evenly and works up rather well, even though it is somewhat brittle and must be handled carefully to prevent creasing and a consequent break. It cures to a pleasing permanent tan with a hard polished surface (flask jacket illustrated in Plate 3 Fig. 2).

Like all vegetable products, the material must be made pliable by soaking in water, then allowing it to lie in a wet towel for a period of time, in this case, overnight.

The strands may be coiled readily if the rods are split and ready for use before they are soaked. This procedure is recommended as the pith can be removed more easily when it is dry. An interesting project might involve unsplit but flattened grain straw (easily done by running the straw through the tightly held thumb and forefinger), made in this fashion.

The bottle over which this jacket was woven is 6¾ inches tall by 4¾ inches at the widest and has a thickness of 1¾ inches. A bottle this size requires strands 24 inches in length for the vertical weavers, split to a width of ⅛ inch and at the least forty-eight strands.

The greater part of the jacket is in simple over-one, under-one weave. The strands are woven together at their centers to form an oblong mat of shape and size to fit the bottle base. Tighten the weave of the completed mat firm and close, holding the weave at three corners with spring clothespins.

Bend the butt of a weaver strand at 12 inches from the end and lay it end up at the fourth corner to form the uneven strand necessary for a continuous weave. Hold it in place with a clothespin and put in a row of weave closely around the mat with its long end.

Catch this strand, turned right side out, into the weave as it reaches this fourth corner, weave across the mat to the first corner, and insert a strand, folded with ends even and the under end turned face up, looping it smoothly over or under the weaver, as necessary, which made the first row of weave around the mat. Adding two weavers at a time does not disturb the rhythm of the weave and allows the filling in of the corner spaces symmetrically. Continue the weave around the mat, placing a pair of weavers at each corner.

If this jacket is to fit tightly, it must be made on the bottle from this point. Tie mat and bottle together firmly with soft string, from the neck over the mat and back to neck. Once the weave has reached the widest point of the bottle, it will be easier to control.

Continue the adding of double strands at the corners on every row of weave until there are enough vertical strands to cover the bottle completely.

Splice weaver strands as needed by slipping the heavy end of the new weaver over the tip end of the old one and carrying both forward as one until the new weaver is fastened securely.

Keep the weaver pushed down in a straight line and the upright strands close together to make a tight weave.

When the bottle reaches its greatest width, the reducing of the number of vertical strands should begin by combining three strands into one at each side center, laying the first and third strands over the second strand and weaving them as one. Combine other strands as the progress of the work and the fit of the jacket demand, placing them symmetrically.

At 4½ inches from the base, allow the weaver to drift to an end on one side and secure the end.

Put in a row of weave around the bottle and fasten the end by weaving it back over itself tightly. This row of weave will be approximately 3 inches from the top of the bottle.

The jacket material will have dried to the point that moisture should be restored so the bottle should be allowed to lie in a wet towel before proceeding with the open weave.

When the strands become pliable, trim the ends of every third upright strand, those under which

the weaver passed, to about 1 inch, turn the ends back over the weaver, and slip them down under themselves so they will lie flat.

The upright strands should now be reduced to nine on each side and to an even number on front and back. The number of strands will vary with the size of the bottle.

Coil a weaver strand and loop it around the three side center strands which are lapped right over left and both over the center. Hold with the left thumb and forefinger, grasp the coiled ends in the right hand with the right forefinger between the coils, hold the work tight between the hands, and give an upward movement of the right hand to turn it palm up.

Lap the right three strands on the side, lay them in the cross, coil the weavers, place the right hand in position, and make the next cross, which will be on the front of the bottle near the side.

This first row of pairing weave, which is put in at least ½ inch above the edge of the check weave, draws in the upright strands by combining them and serves to fasten an open weave decoratively.

Continue the pairing across the front evenly, crossing each two spokes right over left.

When the side is reached, combine nine strands into three groups as was done on the left side and fasten in place with the pairing weave.

Cross the back strands, which are next, the right over the left, and fasten them in place with the row of pairing. This leaves the group of three strands on the first side unwoven. Cross these strands as given, fasten them with the pairing weavers, cross the weavers again and secure them in the row of pairing with the ends under the upright strands.

Examination should show symmetrically shaped vertical crosses and an evenly placed and spaced pairing weave.

Put in a row of pairing close around the base of the neck, lapping right strands over the left and combining the strands on the sides as necessary for a tight fit. Finish off and secure the weaver strands securely.

Put in two other rows of pairing around the neck spaced at ½-inch intervals, but instead of crossing the upright strands, lay them one on the other and weave each pair as one.

Put in two rows of pairing close together around the neck just below the cap base and fasten the ends to withstand the strain of finishing off.

Trim out all surplus strands under each pair just above the pairing weave. Moisten the ends of the remaining strands.

Make a smooth tight coil of each pair of strands, bring it across in front of the pair on its right, and using a large-eyed needle head first carry it behind the third pair of strands and down into the weave on its right through the double rows of pairing at the top and the single row below, to come out between strands. Repeat between * for each pair.

When all strands have been woven into the jacket in this fashion, pull them down neatly to make a small, regular, smooth scallop straight around the bottle neck. Trim the ends about ¼ inch below the row of pairing and push them back under the strands.

Check-Weave Jacket for a Rectangular Bottle.

The clear green glass of the bottle illustrated in Plate 3, Fig. 3., is really too beautiful to be hidden under a closely woven jacket; however, it makes an excellent form for a rectangular jacket some features of which need to be commented upon. The bottle is a duplicate of the one illustrated as Plate 9, Fig. 1, which is jacketed with honeysuckle vine in the Solomon's knot weave. This is a very open weave and allows the color to show through the jacket.

The bottle is 11 inches high, 3½ inches in diameter at the shoulder, and 2½ inches at the base. It is commercially jacketed with flat reed very slightly wider than size medium cane for both spokes and weaver. Cane could be used also for this project with either the finished or the rough side out.

The spokes are arranged on the sides to leave the greater part of the variation in width between base and shoulder to work out on the corners. As there are seventeen spokes, this makes four evenly spaced spokes on three sides with an extra spoke on the fourth side panel. This produces a slightly closer but not noticeably different weave.

All spokes must be kept tight and at a true vertical which, if they are spaced evenly at the base, will result in a very slight spread at the shoulder to take up any extra variation the dimensions of the bottle will cause.

This jacket is another illustration of the versatility of this weave which, with a little planning and the latitude it allows, can be made to jacket practically any form successfully.

As the spokes pass from neck to neck around the base and need ease for angle and finish, the eight spoke strands should be 30 inches long and the extra spoke 15 inches. This will form the extra spoke necessary for a continuous weave. These will all be used at once when the jacket is begun, so they may be coiled together lightly for soaking, but the full-length weaver strands, of which there should be three or more depending on the length, should be coiled individually. This will allow the removal of a strand without disturbing the others in the folds of a towel under which all strands should be placed to become pliable after being soaked in cold water for fifteen minutes. Coiling for the moistening process makes the reed easier to handle and avoids breaking due to folding.

Cross the spoke strands at their centers, three over three.

With a weaver strand, put in one row of weave around the strands as groups, closely.

Slip the end of the extra weaver into this weave at the corner with the end on the underside and fastened firmly for, as a spoke, it will need to be able to resist force.

Spread the spoke groups and weave over one, under one for two rows, maintaining the square shape.

Fold a spoke strand at the middle and slip the end into the weave on the underside at the widest corner space and pull the strand through to the fold. Introducing two strands at a time does not interfere with the sequence of the weave. Fold the other strand at the middle and slip an end into the weave on the underside at the other widest space and pull through to the fold.

Continue the weave, holding to the square shape and making an effort to have the spacing equalized by the time the weave covers the base of the bottle.

Tie the bottle and the woven base firmly together with strong, soft string from neck, over base and back to neck on four sides, being sure the spokes are centered on the sides.

The first row of weave will set the spacing for the spokes and must be given very careful attention. If the base spacing has been properly regulated and the placing on the bottle is true, there will be little difficulty with the spacing, though a constant watch must be kept to see that the spokes do not drift. Pull up on them frequently to hold them at a true vertical and to keep them tight.

Splice the weaver as necessary by placing the new weaver over the old one and weaving both together for 2 inches or so. The end of the new weaver may be trimmed off just beyond a spoke where it will not be noticeable after the jacket has been finished.

When the weave reaches the shoulder, pull up firmly on all spokes and examine the work for tightness and correctness of weave.

Any slant in the shape of the bottle will affect the weave, which is inclined to slip, though an upward or wider slant is easier to control. The side weave will give little trouble if it is put in tight and even and kept pushed down, but the slant above the shoulder toward the neck will require constant vigilance if the weave is to be closely woven. After the shoulder has been well passed, the string ties may be cut out.

The spaces between spokes are so wide that no combining of spokes is required above the shoulder, because they draw together gradually until their edges all but touch around the neck.

Finish off the weave just under the collar made on the bottle about 1 inch below the top and fasten the end of the weaver by running it down into the weave.

Immerse the spoke ends and the neck of the bottle in cold water and wrap in a wet towel for the strands to become thoroughly pliable; reed does not respond well to close-angle manipulation unless it is well moistened.

Bring each spoke end behind the spoke on its right and out on an angle pointing downward. When the loop is pulled down tight these spoke strands will form a slight scallop around the neck somewhat like the edge made on a buttonhole. This is the finish which was used on the jacket and makes an adequate edge which will hold if left untrimmed until the jacket has dried.

A more decorative and firmer edge which would endure washing could be secured by slipping each spoke end back over the spoke on its left, under the two following spokes, and out, continuing the weave from the above point.

If this finish is desired, the spoke ends will have to be slipped back under themselves before the first row of scallops is pulled down tightly.

When all spoke ends have been woven, tighten the first row with a steel knitting needle or something similar and follow with the second row.

Trim closely only after the jacket has dried thoroughly.

Vase Jacket with Shaped Spokes

The distinguishing marks of this jacket are its shaped spokes and applied handles. It is easily identifiable in Plate 3, Fig. 4.

It is made over a thin glass form 6 inches high with a base 2½ inches in diameter and a slightly smaller top. The expansion just above the base gives balance, and the pulling in just above the center to a diameter about equal to that of the base results in a modified hourglass shape.

This commercially made product has served to display attractively and enhance many a small bouquet in its forty years of one-family ownership. The unusual jacket construction makes it worthy of study.

The fifteen spokes are shaped to correspond to the form and the sweet-grass binder is woven around them in over-one, under-one fashion, forming a very attractive lightly ridged surface.

These spokes are made of a very thin soft wood about ½ inch at the widest tapering to ¼ inch at the top and ⅛ inch at the center base. Thinly split oak, ash, bamboo, or any other tough, workable wood, split and sanded, might form the spokes for this project.

Shaping of the spokes would depend on the shape of the form to be used. Much tedious whittling and possible breaking at the center base could be avoided by shaping short spokes ¼ inch at the ends tapering to ⅛ inch at the centers of lengths just short of the base diameter for a spider or base

weave into which the side spokes may be slipped. The over-one, under-one weave will allow this and the fifteen spokes provide the uneven number necessary for a continuous weave. If this method of base construction is chosen, the pointed side spokes should be slipped down into the weave evenly over the base spokes when the weave approaches the edge of the spider. Continue the weave up the sides of the form, pulling the binder strand in firmly to make the spokes hug the form with a tight though not a close weave.

Spoke material and crushable weaver material such as long grass or raffia will need to be soaked in cold water for a short time and allowed to lie in a well-dampened towel to become pliable. This applies also to two pieces of spoke material 9 inches long and ¼ inch wide with ends whittled to a point to be used as bases for the applied handles. The weaver material will moisten more readily than the spoke material so it should be given a shorter dampening period.

Continue the weave to the top of the vase, keeping the spokes true and the weave close, with the rows of weave well pushed together. Fasten the weaver end securely.

Trim off the spoke ends just below the upper edge of the vase.

Wrap sufficient strands of weaver material closely around the upper edge of the vase to make a neat band about ¼ inch wide and fasten it in place with an angled whipstitch of the weaver material. Check for a band of even width placed just below the top edge of the vase, applied tightly with even stitching. Fasten the end neatly and firmly.

Insert the lower end of a handle base into the weave for 1½ inches over one of the spokes, beginning 3½ inches above the bottom of the vase.

Bend the upper end of the base down and insert it into the weave over the same spoke, beginning just under the lower edge of the applied band to make a handle base of about 3½ inches with a nice upper arc.

Fasten strands of weaver material in the lower edge of the band to cover the outer side of the handle base, using the mattress needle. Pull the strands through to make two strands of each. If the best width of the weaver is used, three double strands should be sufficient to cover the outer handle base when they are laid lengthwise on it.

Thread the ends of these covering strands into a large-eyed needle one at a time, and run them down into the weave below the handle as inconspicuously as possible. Bring the ends to the outside so the strands can be adjusted.

Fasten a single strand in the weave close under the top of the handle, then adjust the weaver strands to lie smoothly on the upper side of the handle base and bind them to the base with a firm, angled wrapping from the upper to the lower end of the handle while pulling the covering strands down to fit closely on the base and make a neat, decorative handle. Run the binding strand down into the weave to fasten it.

Take a couple of stitches with the wrapping strand straight across the lower end of the entire handle in line with the jacket weave which will hold the handle firmly and all but hide the stitching.

Make a handle in the same manner on the opposite side of the vase, observing the same measurements so as to secure a matching handle with a like arc at the top. Fasten the binder securely.

Trim off all ends after the jacket has dried.

Over-one, Under-one Jacket of Raffia over Cane Spokes.

In the jacket for the lavender-water bottle shown in Plate 3, Fig. 5, the spokes of size medium cane are planned to fit close together around the neck without combining. The spread of spokes around the larger part of the bottle makes a pleasing variation in the weave.

Fifteen spokes are required for the form 6 inches high by 1 ¼ inches in diameter. As the spokes pass entirely around the bottle, eight strands of cane 16 inches long should be provided. This will be one spoke more than needed so one of the spokes should be woven doubled and the lower part discarded as soon as the end becomes fixed in the weave.

All vegetable products to be manipulated require moistening to make them pliable. Both cane and raffia should be soaked for a short time in cold water and allowed to lie in a damp towel until well moistened.

Splicing of the raffia weaver is accomplished by running the heavy end of the new weaver down under the weave, combining it with the weaver in use, and carrying both along to be woven as one.

Weave into each other at their centers the four pairs of spoke strands, placed one on top of the other, finished side up, so as to make a four-check square. Hold these together with a spring clothespin.

Lay the heavy end of a strand of raffia under this square and slip it into the clothespin.

Begin the weave over and under the spokes around the square, separating the pairs as necessary to provide the uneven number of spokes until all fifteen are in use. Keep the weave pulled in to form a tight base mat just slightly larger than the base of the bottle with a spacing of about ¼ inch between spokes at the edge of the mat.

With a soft strong string, tie this base mat to the bottle on four sides, from neck, over base, and back to neck, being sure that the base mat is centered on the bottle base.

Continue the weave tightly and evenly around the bottle, turning up the spokes to fit closely. If an open weave is to be avoided, the weave must be kept pushed down tightly. This is particularly true on the body of the bottle, for once the shoulder is turned, the weave begins to draw close and it cannot be slipped back to a larger diameter.

The natural drawing in at the shoulder will adjust the length of weave between spokes, but in order to prevent riding up and a consequent loose weave at this point, the weaver must have constant attention and tightening.

The tie strings can now be removed; the jacket should be fixed on the bottle so that they are not necessary.

Continue the weave to make a tight, well-fitting, close weave from the point of the shoulder to the lower side of the cap roll and fasten the weaver temporarily with a rubber band.

Cane is readily broken if turned at a sharp angle, particularly when it is dry, so as the spoke ends are to be turned back under themselves for a finish, the bottle and jacket should be given a short soaking in cold water and then be wrapped in a damp towel for a thorough moistening of the spoke ends which have been cut to about 1 inch and trimmed to a sharp point.

Turn the spokes backward and crease lightly to fit just under the cap roll, beginning with a spoke under which the weaver passes and running the spoke end over it and down into the weave behind itself. Fit it in neatly under the cap roll.

Weave over the next spoke and under the third spoke.

Turn this spoke end over the weaver and run it down into the weave behind itself to fit neatly under the cap roll.

Repeat until all spokes have been woven in.

Finish off by running the weaver down into the weave invisibly and bring it out using a raffia or other large-eyed needle.

Cut the weaver off closely when dry, pulling on the end sharply so that the cut end will not be visible.

Should the center be a little thick for an even base, this may be corrected by moistening it and putting it under a press with a disc at the center. A quarter or half dollar will serve this purpose nicely, and two strips of board for top and bottom, tied together tightly on each side, will make an excellent press.

CHAPTER THREE

THE PALMETTO PALM AND SOME USES FOR ITS FROND

THE OLD SETTLERS found many uses for the sabal or cabbage palmetto frond. They thought it dangerous to be in the sun without a hat so they braided frond and made hats—plain braiding for work hats, fancy-edged ones for Sunday-go-to-meeting hats. Sometimes these braids were closely woven, sometimes very open. There is even a narrow-strand open weave made in the same general fashion as the hexagon weave on the jar shown in Plate 4, Fig. 3.

To combat mosquitoes, they made flails of palmetto with handles trimmed with airily shredded frond. For the ever-present flies, they made really functional flywhisks. The only air conditioning was a breeze, so they made fans of scrub palmetto, bordered with finely shredded cabbage palmetto frond. Hearth brooms and brushes were made from the usually discarded ribs of the leaf or the "fan," as they called it.

They made pincushions, mats, purses, and baskets and shredded the frond to make filling for their bedticks. The mature fans made a rainproof thatched roof when properly put on, and the plant earned its common name because its bud leaf could be cut out and boiled when there was no cabbage in the garden. Taking the bud leaf destroys the palm, but that was of no consideration when the swamps and much of the dry land were covered with them.

Today, commercial harvesting for Palm Sunday use brings in a tidy sum to a few, who begin cutting the leaves early in the spring as they reach the right stage of development, and stacking them like cordwood under a tarpaulin until the time comes to ship them.

As the palm has thirteen new leaves a year, roughly one every moon, the taking of a few of these secondary leaves does not injure the plant, so it is a continuing crop. The leaf has to be harvested when it is ready and at a certain time of the moon, for if it emerges from the plant during the full moon, the light both night and day will usually force it to begin opening before its growth is completed, and the chlorophyll will develop and color it.

When it is harvested at the right time, the leaf will have a 3-inch or longer stem above the crown of the plant, tightly folded and sword shaped. This is easily pulled out and downward so the stem can be readily cut, even with a pocket knife, at the point of the bend where the fibers are taut.

Taken at the right time and cured under the proper conditions, the frond is a creamy white, durable, easily handled, double-faced, washable material with which it is a pleasure to work.

Directions for harvesting and curing the frond will be found in Chapter 12.

Open Mesh Jacket Bound with Pairing Weave

This weave is highly adaptable, decorative, and not too difficult even though it is made with palmetto frond stripped to a width of little more than 1/16 inch.

In addition to the over-one, under-one weave used in examples already given, this jacket uses the pairing weave in the open work on the sides. Directions for the pairing weave are given in Chapter 12.

The jar shown in Plate 4, Fig. 1 is 5 inches high, 4½ inches in diameter at the shoulder, and top and base diameter of 2⅝ inches. Fifty-two strands are required for the 104 spokes, as the strands pass entirely around the jar, making a strand length of at least 27 inches necessary. Ten or more evenly stripped full-length strands of the same width should be provided for the pairing weave and the check weaves at the base and top.

As with all the examples, an exact duplicate of the jar illustrated may not be possible to obtain, but these figures will give some idea of the material necessary and the adaptations required for the form at hand. For a larger form, the instructions could be adapted to the use of wider or more numerous strands or to other mediums such as cattail rush, either split or coiled, commercial chair cane, or honeysuckle or other vines which are pliable when moistened.

Weave together tightly at their centers 24 strands in over-one, under-one or check weave. The strands used for this jacket were so narrow and light in weight that they were very unruly; however, after a couple of cross strands were woven in and the weave clipped with spring clothespins, the work became more stable. Weave in all strands, twelve into twelve to make a square, pulling together to make a firm, tight, square weave.

With a long strand, the butt end of which is allowed to extend for 4 inches on the left, begin at a corner and weave around the square in check weave. A firm tension should be kept on the weaver while it is being put in. This will tend to round the square at the corners but it can be brought back into shape by tightening the strands with which it was made. This will be necessary anyway for a tightly woven base weave.

A continuous check weave requires an uneven number of spoke strands and explains the 4-inch extension of the first weaver, which was held in place with a clothespin until it could be used as a spoke in the weave.

Proceed with the check weave around all sides of the square, pulling on the strand ends frequently to shape and tighten the weave.

The layout now should have the shape of a cross centered by a woven square. The open corners must be filled in to cover the base of the jar and to furnish the number of spokes required for the sides of the jacket. This is accomplished by inserting spoke strands, a pair at a time, into the open corners. Fold a new strand on an angle so that it will fit into the corner and can be woven in sequence as the weaver reaches that point. Two strands inserted at one time do not affect the sequence of the weave, so the check weave continues without disruption and helps fasten the new spoke pairs which must always be lapped in pattern, that is, left over right or right over left, according to the fold in the first pair inserted.

Continue to insert strands in this manner until there are seven inserted pairs at each corner or until it is filled.

Extension of the inserting of strands in the outer base weave is entirely acceptable, and decorative, if the fold pattern is maintained. Should the spaces at the corners be too small for as many as seven pairs, those not needed may be dropped.

When the weave becomes large enough to cover the base of the form, fasten it firmly to the form from top, over the base, and back to the top of the form on four sides with a tight, strong string.

Continue the check weave around the outer base of the jar for four rows, tightening the spokes and the weaver frequently for a very close weave.

When the fourth full row has been woven and the weaver has reached the beginning point, slip the weaver back over itself closely for about 2 inches, thus disposing of the end. Trim closely only when the work has dried thoroughly.

Turn the short extra spoke end back over the last row of weave, slip the end down into the weave over itself, and pull tight. Adjust the spokes to take up the space left by its removal.

To make the open weave, lap the strands which made up the check-weave base left over right to form spoke pairs and place a rubber band just above the cross.

Go over the work to see that the spokes rise at a symmetrical angle and that they lap left over right.

At a space of ⅝ inch up on the spokes, the outer left spoke of one pair will fall naturally over the outer right spoke of the next right pair of spokes. Loop a weaver strand for pairing around such a cross and put in a close, symmetrical pairing row around the jar 1½ inches above the jar base, weaving around each pair of crossed spokes in turn.

When the weave reaches its beginning, coil a weaver end lightly (it breaks less readily when coiled), slip it into the loop made around the cross, and follow the pairing row back for 1 inch or more. Moisten the other weaver end and weave it back over the pairing in the opposite direction as invisibly as possible.

Cut off the excess but leave close trimming until the work has dried.

Check to see that the pairing is put in properly with no reverse weaves and no extra turns between spoke pairs. Make sure that all spokes lap left over right and that they hold the original angle. The spokes will fall naturally into a cross of left over right but careful checking is necessary to see that there is no accidental disarrangement of sequence.

Distance between rows of pairing is a matter of preference controlled in some measure by the size of the form. The pairing in the jacket illustrated allowed ⅝ inch between rows, appropriate for the form.

Four rows of pairing and the open weave between them will cover the side of the jar to about ⅜ inch below the point of the shoulder, combining two of the spokes unobtrusively so as to secure the necessary uneven number and crossing the left spoke over the right above the shoulder as the weave is being put in.

Make no further spoke crosses but fill in the space up to the cap rim with check weave, pulling the spokes tightly together. The base and the neck are the same diameter, so there should be no difficulty in fitting in all spokes.

The form used has a rim for a friction top cap under which a stay strand, preferably a very narrow outer edge from the frond or a very narrow strand of mulberry root, should be tied tightly in a square knot which will not slip. Either must be made pliable by moistening.

If the spoke ends have become dry, moisten them well by wrapping the jar in a wet towel. When the ends are pliable, turn them over the stay and weave them down over themselves into the weave for about 1 inch, spreading the ends of the tie strand and placing them so that they are covered with the turned-over spoke ends.

Do not trim closely until the jacket has had time to dry.

PLATE 4. Palmetto Frond Jackets:
Open mesh jacket with pairing weave binder (fig. 1); diagonal check weave on an olive bottle vase (fig. 2); all-over hexagon mesh jacket with a woven cap cover (fig. 3); check weave jacket for a three-cornered bottle (fig. 4); diagonal check weave jacket for a round jar (fig. 5).

Diagonal Weave for a Straight Bottle

Plate 4, Fig. 2 shows an ordinary olive bottle 5 ½ inches high with a 4-inch circumference.

The jacket is made of palmetto frond cut to measure (very important, giving more length), rather than stripped, and must be taken from the longest and best fronds so as to avoid splicing. The leaf and hence the stripped frond taper naturally; and a sufficiently long strand with a constant width can be had only by cutting. The strands should be at least 5 inches longer than double the height of the bottle, since they must be folded in half to make the so-called buttons with which the weave is begun, and to allow for the spiral weave and for the finish.

This is a simple, closely woven check weave, but the method of starting the weave by the connected buttons, as the double V forms were called by old-time palmetto workers, the turning over the base strands in rotation to give the spiral line, and the smooth turn of strands to fit the side of the bottle are features which may be adapted to other jackets if the worker will master the procedure.

The size of the bottle regulates the number of V pairs or multiples of four and the width of strand. For this size bottle, a strand width of ¼ inch seemed most suitable. Wider strands may be used if the form is large and a bold effect is wanted, but the strands would have to be very long ones, woven together at the butt ends to take advantage of the entire length. The turning over of the strands at the edge of the base would give the spiral angle to the weave.

Palmetto requires dampening before it is manipulated. The prepared material should be soaked in cold water for fifteen minutes and allowed to lie in a wet towel until it becomes pliable. Should the work dry out or have to be put aside, repeated wetting will be necessary. This may be done safely unless it is allowed to lie in the towel until it sours and turns it a permanent yellow. Cutting the frond to size before it is moistened is advisable and contributes to accuracy of measurement as it swells slightly in the moistening process.

Make a double V by folding a strand in the middle from the right. Lay another strand with its center in the crotch of the V. Fold the right end of the second strand over the right side of the V to lie parallel with the left side of the V. Fold again from the right to weave over one, under one to make a VV or button. Crease folds firmly and hold with a spring clothespin.

Repeat between * to make three other double V forms.

If the strands are manipulated in the order given, the VVs will be alike on back and front so the following arrangement of buttons may be made without regard for a certain face. Place the buttons two and two, ends touching, and run a

soft twine or, preferably, the soft outer rib of a palmetto frond stripped to 1/16 inch through the very tip of each V and tie in a square knot. When the rib is thoroughly pliable it ties and holds firmly if a square knot is used. This results in four groups of four strands each in which the two right and the two left strands are turned at sharp angles away from each other around the tied center.

Hold this round center with its extending strands in the left hand, turn over to the outside each left outside V strand, and weave into the adjacent left VV under one, over one. When all four left outer strands are woven and the clothespins reset, check to see that the original groups have four strands in each group throughout the weaving of the base. Should any group be found with an uneven number, either some step has not been taken or some strand has been woven out of turn.

Turn over to the outside each of the eight outside V strands and weave into the adjacent right VV under one, over one.

Turn over to the outside and weave into the adjacent left group all inside, or lower, left strands under one, over one.

Turn over to the outside and weave into the adjacent right strands all inside, or lower, right strands under one, over one. This turning over of the strands allows the weave to conform to the round shape and to lie smoothly on the base. It also sets the angle for the spiral weave on the side of the jacket.

As soon as the woven base will cover the base of the bottle, place the bottle on the base and begin the weaving of the sides by taking any strand and weaving it into the adjacent strands as far as strands on the opposite side present themselves. Tighten the weave by pulling up on the strands and reset the pins. If the weave is kept tight, the clothespins will soon be unnecessary.

Weave the strands in rotation until the sides are closely covered and the weave extends about ¾ inch above the edge of the bottle. Tighten the weave and straighten the upper line of weave.

* Make the points for the collar by folding the first and fourth strands to the front on a straight line across themselves where they cross above the intersection of the second and third strands. Hold firmly in place. Fold the end of each of these strands back over the lower opposite strand of the pair and on itself into the weave to make a flat triangular point. *

Repeat between * to make a neat finish for the other strands and form a line of connected points around the jacket at a right angle to the vase.

Press the points into position by placing the vase in its well-dampened jacket, top down, on a flat surface with the points spread and a weight on the base. Fit with a small, shallow base such as the turned orangewood base 3⅛ inches in diameter shown in the illustration.

Trim strands closely at the edge of a check when the jacket has dried.

Hexagon Mesh with Palmetto Frond

At first sight the weave on the jar in Plate 4, Fig. 3, seems closely akin to cane chair seats. Actually it is much less intricate and less time-consuming. The weave is a matter of laying in a horizontal weaver strand at accurately positioned spoke crossings. Unlike the miniature jar jacket in the same weave shown in Plate 14, Fig. 3, the spoke strands end at the cap base, producing a much less involved finish, and the base is woven in a simple over one, under one instead of in a hexagon mesh.

The cap cover is time-consuming; its very fine weavers make it progress very slowly. Though it is an interesting project and makes an attractive finish for the jar, it may be omitted.

The medium usually used for this weave is palmetto frond, usually readily available in an area where the cabbage palmetto grows. Other very thin, pliable material which will split to a constant width may also be used.

The weave, which must always have an even number of spokes, adapts to any straight-sided, round, or rectangular form. The allover mesh of three rows will form a diamond with the opening in row three directly above that in row one, in which case the weave is regular and symmetrical with rows of openings at an even distance from the base; this is the weave that must be used with strands of comparatively short length. Men's toilet goods frequently come in commercial jackets in which the mesh is on an angle and the jacket woven throughout in a hexagon mesh. This is done by wrapping the warp strands which make the X forms at opposite angles around the bottle with not more than two strands of material. The cross weaver is the usual horizontal strand which finishes each row independently.

The development shown in Plate 4 is for a jar 4 inches high with a 4-inch diameter and requires eighteen ⅛-inch-wide strands of palmetto frond 16 inches long. Four times the height of the jar may seem a very liberal allowance, but it is needed for the base weave, the angled spoke strand, and ease to make the turned-over finish. Also, should a strand break, the end could possibly be pulled through to furnish sufficient length on both sides without having to replace the strand. Strands should also be provided for the horizontal weavers and for the base weavers which should be left at full length.

In planning for the jacket, the size of the form must be taken into consideration as well as the width of strand and the size of the mesh. This will readily be seen by a comparison of the two examples of this weave.

Soak the jacket material for fifteen minutes and wrap it in a wet towel to make it pliable.

Make two groups of three spoke strands each, placing them on each other and cross them at a right angle at their centers, then on a cross angle.

With the lower end of a weaver strand, the end of which has been allowed to extend for about 6 inches on the left to form the uneven number of spokes necessary for a continuous weave, cross these spoke strands from both angles and weave them as groups for two rows, pulling the weaver in tightly.

Spread the spoke strands in the next two rows, weaving over one, under one. Splice by laying the new weaver over the old and weaving them as one.

Fold a spoke strand in the middle, insert one end under the weaver at one of the widest corner spaces, and pull the end through to the fold. Two strands inserted at once will not disturb the sequence of the weave.

Proceed with the weave, inserting the other spoke strands in this manner until all eighteen have been woven in and the base weave will reach to within ¼ inch of the outer edge of the jar base. Work toward an even spacing of the spokes as soon as all strands are inserted.

Fasten off the weaver and the extra spoke by running them back into the base weave when it reaches the proper diameter. Pull firmly on all spokes to tighten them. The X forms will be more regular if the last row of base weave is treated as a horizontal weaver and the spoke strands are regulated so that the right arm of the X has the over weaver on its base. Examine the placing to see that the crosses are equidistant.

Tie the woven base and the jar firmly together from top, over base and back to top on four sides, centering the base on the jar carefully.

Cross the right strand over the left, laying the right strand against the jar and bringing the end of the left strand back toward the worker so a horizontal weaver may be laid in the upper arms of the X or under the under arm and over the over arm.

Lap this horizontal weaver for 2 inches or more, leaving a small extension of it for tightening.

This first horizontal will be very near the outer base of the jar and should be placed at a true right angle to the center of the Xs.

Hold the weave in place with a heavy rubber band above the weaver.

The strand ends require rearranging after each horizontal weaver is put in. While this is being done, pull up firmly on the spoke strands, taking care to have the X forms at a true vertical and the spoke strands rising at a symmetrical angle. A good test of this is the diamond formed by the three first rows of hexagon openings. Once the angle of the strands is established, it is easy to hold in place.

The sixth horizontal will bring the weave to within ½ inch of the depression in the jar below the cap base which will serve to hold in place a tie around the strands. This, preferably, is a very narrow rib section from the outer edge of the double frond, not the midrib which is stiff and coarse. The outer rib may be only about 1/32 inch wide but is very strong when thoroughly moistened and is the same color as the frond.

Soak the work in cold water again and wrap the jacket and jar in a wet towel for the strands to become pliable.

Wrap the palmetto rib tightly around the strands twice at the depression below the upper edge of the jar, or at the cap seat if there is no depression, and tie it tightly in a square knot which will not slip.

Pull up firmly on the spoke ends nd bring right and left strands from adjacent pairs together so they will fold evenly, one over the other, at the tie line. Bring these folded strand ends to the front in an inverted V form and weave them down, each over its opposite, to at least the second horizontal weaver row.

Repeat between * to finish off all strand ends.

A strand may be run through the top turns to cover the tie or it may be left exposed if the tie is a palmetto rib. If string is used, it will require a covering strand.

Woven Cap for a Jar Cover

A woven cap for a jacketed jar, especially for a large one, gives a pleasing finish and enhances the overall appearance.

The center section of the woven cap in Plate 4, Fig. 3, was made with four spoke strands of palmetto 7 inches long and ⅛ inch wide which were bound in the center with another strand 4½ inches long. This extra strand with its binding allowance makes the uneven number of spokes necessary for a continuous weave, in this case, nine spokes.

With a weaver 1/16 inch wide, put in three rows of over-one, under-one weave. Fasten temporarily with a doll clothespin or a paperclip and pull the spokes into an even spacing.

Proceed with the weaving, making the strands lie very close together. Splice weaver strands as necessary by laying the new weaver over the old one and weaving them together.

When the weave has reached a 2-inch diameter, the center circle has been woven. The narrow strands are inclined to lie on edge rather than flat; this is encouraged and results in the raised center. In the rest of the weave, the strands are made to lie flat, which accounts for a different appearance with the same weaver width.

Slip into the weave from the back, about ⅛ inch from the outer edge of the center weave, a 4-inch

spoke strand and pull it through to its center to form a pair of spokes between each two spokes. Adding two spokes at once does not disturb the sequence of the weave.

Weave a tight over one, under one between evenly spaced spokes until the weave will cover the jar top, place the cap on the top, and hold it in place with rubber bands.

With a long weaver of the same width as that used for the horizontal weaver in the jar jacket trimmed to a long slope, begin a right-angle edge or collar for the cap by putting in two rows of weave while pulling the spokes down closely and evenly.

Turn the well-moistened spoke ends back on themselves tightly so as to fit the depth of the cover, pushing them up through the weave to the top of the edge, and leave them for later close trimming.

Put in as many weavers as necessary to fill in the collar space except one row which is to be woven when a well-moistened soft palmetto rib has been woven into the edge, pulled tight, and tied in a square knot. Pass the spread ends neatly back into the weave.

The trimming of the lower edge of the top weaver will allow the last row of weave to be turned back on itself to make a straight lower edge. Pull the strand tight but leave close trimming until the jacket has dried.

Check-weave Jacket for a Three-cornered Bottle

Other jackets in this chapter are woven partially in simple check weave or two with a diagonal check; this is the only example in this group made entirely in the over-one, under-one procedure. It shows how attractive a simple weave can be, and gives the method of handling the difficulties presented by a three-cornered bottle (see Plate 4, Fig. 4).

The bottle itself is small, only 6 inches tall with 2-inch side panels, so the frond for the jacket was stripped to ⅛ inch for both spokes and weavers.

Each side required fifteen spoke strands, so twenty-two strands 18 inches long were needed for the side strands which pass entirely around the bottle and the seven spokes which form the weaver strands for the base and the spokes for the third side. Most jackets are made with a base in which the spokes are fastened together with a weaver, but this base is made by weaving the spokes for one side into those for the other sides.

Full-length strands will be needed for the weavers, of which eight or more should be provided. These will need to be spliced by laying the new strand over the tip of the old one a few checks before it is needed, carrying them both along for about ¾ inch.

Soak the stripped frond in cold water for fifteen minutes and wrap in a wet towel to make it pliable.

At the center of three spoke strands laid parallel, weave in a spoke strand over one, under one, over one, and pull the strand through to its center.

Turn the upper end of the strand under and weave it under one, over one, under one to make an inverted V form. These weaver strands should be inserted alternately on right and left of the V to maintain a balance. Hold weaver with spring clothespins or paperclips.

On the left side of the V, weave in another strand over one, under one, over one, under an inserted side spoke strand at its center, and over another inserted strand. Pull the V strand through to its center and weave it down the right side of the inverted V under one, over one, under one, over one, under one. Pull all V ends to tighten and reset the clothespins. Note that there are two inserted side spoke strands above each inverted V tip.

Follow this routine until all of the fifteen side spoke strands and the seven strands which have been used to make the V weavers have been woven together at their centers to form the base of the jacket.

For a continuous weave, an uneven number of spoke strands is necessary. Provide the extra strand by allowing the butt end of the first weaver to extend for the length of the other weavers beyond the tip of the V and weave it down the side of the V in sequence. This will also provide the fifteenth spoke strand for the third side.

Continue the weave around the triangular base in the proper sequence, pulling tightly on the weaver. Hold with a clothespin and examine the work to see that there are no errors in the weave and that the spoke strands are tight.

Set the bottle on the base, turn the side spokes up to lie smoothly around the bottle, and put in a few rows of tight weave, keeping the spokes close together and at a true vertical.

Should there be insufficient spokes to cover the sides of the bottle, this may be corrected by weaving in a strand the same length as the spokes at one or all corners. Remove the bottle and weave from the top of the weave to the edge of the base, pull the strand to its center, turn it and weave from the base to the top of the weave. Two spokes added at one time do not affect the weave sequence.

Keep the weave tight and the spokes true with the weave centered on the base of the bottle. It is difficult to tie the base weave to the triangular shape and as the angles of the bottle will help hold the weave in position, little more than frequent examination and correction will be necessary until the progress of the work will hold the jacket firmly in place.

Continue the weave around the sides tightly

with rows of weave placed close together and tight, true-angled spokes.

When the shoulder has been reached and there is no space for another row of weave on the sides, put in a row just above the shoulder.

In the next row of weave, bring two spokes together under a corner spoke and weave the three as one spoke. Since two spokes are removed, the weave sequence is not disturbed. Combine the three strands on the other two corners in the same manner and in the side centers. The narrowing of the bottle above the shoulder will require the combining of strands at corners and center sides two or more times which may be done symmetrically with a little planning. Carry the ends along under the weave until it is sufficiently advanced to secure them; they may then be cut off. By the time the weave reaches the base of the neck, only enough strands should be left to cover it closely.

Make two rows of weave close around the lower neck. Hold with a clip.

Turn the spoke ends back over themselves, under the weave of these two rows and to the outside, leaving a loop ½ inch or so above the top of the bottle. Fold back all strands in this manner.

Proceed with the neck weave closely, treating the double spokes as if they were one. Pull the weave in closely under the roll around the top of the bottle and fasten the weaver by running it back under the weave.

Pull the loops down so they will fit evenly under the neck roll.

Trim all ends when the jacket has dried thoroughly.

Diagonal-weave Jacket for a Round Jar

Though the method of weave for this jacket is much the same as the straight bottle vase in Plate 4, Fig. 2, the directions for a different base weave are given as the enlarged midsection of the jar requires extra strands to cover the surface. These are provided by making the base strands double, laid one over the other and woven as one.

The material used for this jacket is stripped palmetto frond, instructions for the handling of which will be found in Chapter 12, though any flat, pliable fiber without definite face, or grain straw if it is flattened but not split, could be used. However, grain straw is not recommended for the beginner on account of the breaking hazard.

The jar in Plate 4, Fig. 5, is 4 inches high, 3½ inches at widest diameter, with a 2-inch base. These measurements are not mandatory, and adaptations can be made for jars of other sizes. More strands or wider ones are permissible though the narrower strands are much easier to handle.

From the longest and best fronds, prepare thirty-two strands of palmetto frond stripped to a width of 3/16 inch. A constant width for at least six times the height of the jar should be allowed to cover the base, take care of the diagonal side weave, and give ease for finishing. Measure from the butt end and discard overlong narrowed tips of the frond. Prepare also several of the longest outer ribs from the leaf frond, not the midrib, which have been stripped to a width of 1/16 inch or less, to be used as stay strands. These are very strong and tie with ease when they are thoroughly pliable. Also, they are the same color as the weavers.

Soak the prepared material in cold water for fifteen minutes and wrap in a wet towel until it is pliable.

Pair four strands and lay two double strands across the other two at their centers on a right angle. A more even weave will be secured if the strands are laid butt to tip when they are being paired for the base weave.

Follow with other double strands woven on the sides of this center until all strands have been used and the weave measures 2 inches square. Hold with spring clothespins.

Put in a row of pairing weave (see Chapter 12) tightly around the woven square to hold it, and remove the clothespins. This row of pairing is necessary to tighten and hold the weave for the base firmly. It may be of string, to be cut out later, or it may be of the prepared outer palmetto rib, which shows very little and may be left in the weave if preferred.

The strands should be lapped evenly and centered when woven. Examine the work and make any necessary adjustments before tying it to the base of the jar in the following manner.

Wrap a string twice around the jar neck and tie it firmly. This will act as a stay to which the end of another piece of string can be tied, passed down over the woven square on the jar base, and back to the neck to be tied tightly around the string.

Repeat with a string tied on the other half of the jar after centering the woven square carefully on the jar base.

The angle of the diagonal weave is set in the first few rows above the square base by turning the strands in the side centers of the square over on an angle away from each other to lie underside up. At least two strands will need to be turned for the sides and they will be on the base, but they give the needed direction for the strand angle.

There are more turns at the corners of the square, however. These pull in the weave around the lower base of the jar and make a small pattern for about 1 inch up the sides, equidistant around the jar. These strand turns begin at the edge of the pairing, turn over at an angle to the inside, and lie face down, one lapping over the other. The order of lapping depends on the order of the weave and may be right over left or left over right, but it must be consistent throughout the weave. Should a double weave be necessary, that is, two strands over or under one, make it on the base so that the weave is regulated before the weave on the side is begun. If the proper sequence of weave is established, there will be little difficulty in maintaining it or recognizing the order of the weave. The turning of the strands is a necessary device to tighten the weave, give direction to the weavers, and fill in the space. It also makes a more decorative weave and accommodates itself to the shape of the form as the vertical-strand weaves could never do.

The over-one, under-one weave is used throughout with tight strands well pushed together to fill in closely the space at the base of the neck.

Splicing, if necessary, is done by slipping the butt of the new strand over the one to be spliced about three checks before it is needed. The old strand should be trimmed off after about 1 inch but the short end of the new strand should be left for the final trimming.

When the base of the neck has been reached, remove the strings which tie the base weave and the jar together. Hold the weave at the neck with a strong rubber band and dampen the ends of the strands for turning. The frond breaks at close turns when it is dry.

Tie a well-moistened palmetto rib around the strands at the base of the neck by wrapping it around twice and tying it tightly in a square knot. This will stay the weaver strands at the base of the neck and make a neat finish. Pull up on all strands to take out any looseness.

Pair the weavers where they cross and weave them back on themselves for a few checks. This will not be noticeable if the weave has been made tight so that the jar is well covered and the right strands are paired.

Such a finish will leave the glass neck of the jar exposed but this is not unsightly. However, if a finish like the one illustrated is desired, it may be done with a braided palmetto band the width of the neck, fastened invisibly to the jacket at the base of the neck by a very narrow frond rib run alternately through braid and jacket edge folds.

The braid shown is made of six strands of frond ⅛ inch wide. The rule that all braids made with strands of an even number turn one edge under, one edge over makes the braiding of such a collar simple. If the procedure is practiced until a tight, smooth-edged braid can be made, if the collar is drawn together tightly around the neck and the end woven back over the braid or even turned under neatly, it adds a decorative finish.

Dry thoroughly with the base placed so that the air can get to it, and trim all ends closely.

CHAPTER FOUR

CATTAIL-RUSH PROCEDURES

THE COMMON CATTAIL rush is one of a very large family of rushes which grow in many parts of the world and which are used in various ways from mats and rugs, both braided and woven, to chair seats.

It is found frequently growing in water at the edges of ponds and boggy places, sometimes growing to a length of 6 feet or more when conditions are favorable. It should be harvested when the leaves are at their best, usually when the tips begin to turn brown in the late summer. The underwater part of the leaf will be a different color and texture so it should be cut off, though not discarded, for even if the length limits its use, it works up very well (see the small jacket made with a square knot mesh in Plate 5, Fig. 2).

Splice butt to tip for a more even weaver. This will necessitate the removal of a too-heavy butt. When the tip becomes too thin to maintain the coil size, wrap it in the butt of a new leaf to continue the coil. The cattail leaf is double-faced, so little attention needs to be given to keeping one side uppermost.

Cattail leaves are usually used whole and either coiled or crushed; however, they make a beautiful check weave in the flat when closely woven of leaves carefully selected for size. They may be woven in an upright check to make a cover for a rectangular form, or woven diagonally on a straight-sided round bottle if the size will take a very bold weave. Such jackets could be made by adapting directions given in Chapter 2 on check weaves or in Chapter 3 on jacketing a straight bottle with a diagonal weave. For harvesting and curing cattail leaves, see Chapter 12. Directions for the pairing weave are also given in that chapter.

Jacket for Grandmother's Vinegar Jug

The appeal of this jug is in its shape, enhanced by the soft grey sheen of its glaze. (Plate 5, Fig. 1). It had seen many years of service in New York in the days when housewives manufactured their own vinegar. Somehow it survived the vicissitudes of many moves and its owner, a Florida friend, gave it up only when she knew it would be jacketed.

Soft grey-green rush was chosen for the jacket because it complements so beautifully the glaze, jug and rush seeming to have an affinity. Cattail is also practical for the purpose because of the length of leaf.

For irregularly shaped forms, spokes should be long enough to pass entirely around the form from neck over base and back to neck with an allowance for enlargement, angle, and ease in finishing off.

This jug is 10½ inches tall and 26 inches at largest circumference, tapering to a base circumference of 16 inches. The spokes do not run straight up the sides, so an allowance of at least 8 inches on each end of the neck-to-neck measurement of 30 inches, or a spoke length of 46 inches, seems desirable.

Prepare and moisten twelve leaves for the twenty-four spokes and ten or more for the weavers, as directed in the beginning of the chapter. It will be wise to wrap spokes and weaver leaves in different groups, as there may be variation in size or length of the two groups.

To make the base for the jacket, weave into each other at the centers of the flat spoke leaves, alternating butt and tip in order to have a more even finish, the twelve selected leaves in simple over-one, under-one weave. This gives twenty-four ends for the spokes, six on each side of the square.

Pull tight and fasten three of the corners temporarily with spring clothespins.

Near the fourth corner, loop a lightly twisted cattail weaver around one of the spokes off center, so as to provide for spaced splicing, and begin the pairing weave on the mat which covers the bottom of the jug. Directions for the pairing weave, given in Chapter 12, should be mastered before the jacket is begun.

The work proceeds from left to right which means that the woven base cannot be turned over and the jug must be set on the upper side of it when the work reaches the place the spokes must be turned up around the form. Bear in mind that any exposed end or other irregularity in the weave should be on the upper rather than on the lower or outer side of the mat.

In the case of the base mat, the weave is put in over a single spoke, the aim being to make a firm round mat. The check-weave square can be rounded by pulling firmly on both weavers and spokes at the corners. This will crush the spoke strands and will contribute to the firmness of the base weave. Continue the weave until a tight, symmetrical mat the size of the bottom of the jug has been built up. The pairing turns between spokes will lengthen as the circle expands. This is normal and contributes to a closely placed weave. The entire jacket should be a long, easy coil with just enough twisting of the weavers between spokes.

When the weave will cover the bottom of the jug and the pairing reaches the spoke around which the weave was begun, put in another row of pairing close beside the last, crossing every third spoke over the one next to it and proceeding as if the two were one. The spokes may be lapped left over right or right over left but when the routine is established it must be maintained both for appearance and for ease in tightening the spokes.

Set the jug on this woven mat, centering it carefully, and put in three more rows of pairing closely around the outer base of the jug, pulling upward on the spokes to make a snug fit. When the third row of pairing around the outer base has reached the spoke where the weave was begun, the rows of pairing should be the same and measure just under 1 inch in width.

Secure the ends of the weavers by running them down into the weave one at a time, invisibly, and out on the inside of the woven base. A large-eyed needle, preferably a bag needle, is a necessity for such work as this. If it is used eye first, there will be little likelihood of splitting spoke or weaver where it is inserted. It is also helpful in replacing a broken spoke as it will carry a new spoke back along the weave until the end can be fastened firmly in the base when the damaged spoke can be cut and withdrawn. The spokes in the base mat were crushed by a tight weave but the greater part of the jacket is very open and the spokes will be more regular if they are coiled very lightly before they are crossed.

The spokes have been given direction by crossing them in the first row of outside pairing. The three subsequent rows have maintained the angle, and will cross again about 1 inch above the last row of pairing around the outer base of the jug. The worker must adjust them symmetrically, to be sure they are crossed in the chosen order and kept taut. To maintain the pattern, each row of pairing is finished independently, as a closed ring not connected with any other row.

Before proceeding with the open weave of the sides, tie the jacket base firmly in place on the jug, with centers even, by running a soft string from the neck, through each side of the base, and back to neck. Repeat on the opposite side.

Fix the pairing strands in place by looping a lightly twisted leaf, off center for the sake of splicing, around two spokes at the point of cross and continue around the jug, crossing the weavers once between each pair of crossed spokes and once around them. When the row of pairing reaches the point of beginning, tighten the weave and run one of the weavers through the loop made around the pair of spokes at the beginning of the row of weave. Pull tight and weave neatly backward into the row. Weave the other strand forward into the coil, lapping for two or three crosses, but do not

PLATE 5. Cat-tail Rush Jackets:
Old jug jacketed with coiled cat-tail, with pairing weave binder (fig. 1); underwater cat-tail jacket made with square knots (fig. 2); variation of vinegar jug weave (fig. 3); cat-tail rush jacket with rush spokes (fig. 4).

cut the weaver strands off short as the weave may require adjusting later and a manageable end will prove handy. Much tightening or adjusting will be avoided by placing the weaver ring at the natural spoke cross and maintaining the position a measured distance from the base.

The spread of the jug will fix the approximate place of spoke cross and ring of pairing weave; the spread between crosses will regulate the increase in pairing turns.

Follow with rows of pairing as given for this first row until all are in except the top one near the base of the handle. The string ties are now no longer needed and may be cut.

The top row is put in the same way as the other rows, except that the decrease in the size of the jug will require fewer turns of the weaver strands between spokes. After removal of the stay strings, the jug can be turned so that there is a pair of spoke crosses on each side of the handle.

Reduce the number of spokes by twisting the arms of a pair independently, then very lightly together, and fasten them with a row of pairing just below the base of the neck. This is the first of four continuous rows of pairing around the neck of the jug. Hold the weavers with a spring clothespin and inspect the work thoroughly, tightening the spokes and making any necessary adjustments or corrections in the weave.

When the pairing collar has been put in, finish it off neatly by running the weaver strands back into the weave invisibly.

Moisten spoke ends and collar thoroughly.

Make a row of tight scallops around the upper edge of the woven collar by twisting a pair of spoke ends into a tight coil, thread them into the needle, bring them around the outside of the pair of spokes next on the right and down into the weave behind the third pair of spokes. Use the bag needle eye first and bring the ends out below the third row of collar pairing counting from the top.

When all spoke pairs are woven in, pull the ends down gradually until a smooth tight scallop edges the weave just below the top of the jug.

Allow to dry and cut off all ends closely.

Jacket in Square-knot Mesh

The brilliant green glass of this bottle contrasts nicely with the shiny tan of the jacketing material to make an attractive combination.

For such color and texture, the material, cattail rush, must grow under water. The depth of the water in which the plant grows determines its length, and it is always short because its preferred location is the edges of ponds or small streams. The length of the jacketing material limits the size of the bottle that can be used.

The bottle illustrated was common on grocery store shelves fifteen or so years ago. It is of an es-pecially pleasing color and dimension, measuring 4 ½ by 3 ¼ by 1 ½ inches (see Plate 5, Fig. 2). With the round braid loop it can be a hanging plant container.

This underwater part of the cattail leaf requires moistening, although it absorbs moisture much more readily and is easier to handle than the upper part of the leaf. It splits true and ties readily, making it a good choice for this jacket.

Splicing is accomplished by placing the new weaver, edges trimmed lightly if necessary, in the old one, coiling them lightly, and proceeding as if the two were one.

Prepare the material by taking off the thin outer edge which grows on each leaf butt. This is nice fiber but it widens at the base which makes it unsuitable. Soak these leaf ends, which should be at least 12 inches long for a bottle of the dimensions given, in cold water for about five minutes and wrap them in a wet towel to moisten them thoroughly.

Split the ends into strands ¼ inch wide and select eight of the longest for the sixteen spokes. The strands are woven together at their centers to form the base mat and furnish the spokes on all sides.

To cover the base of the bottle, an oblong 2 by 1 ¾ inches, three strands were woven into the other five in check or over-one, under-one weave, tightened, and held in place with spring clothespins at the corners.

Split one of the longest leaf ends into ¼-inch weavers.

Coil one of these lightly and loop it around one of the corner spokes, off center so that splicing will come at unequal spacing, and put in rows of pairing (see Chapter 12) to fill out the edges and make a closely woven firm mat to fit the base of the bottle.

Center this mat on the bottle base and tie it tightly in place with soft, strong string from neck, over mat, and back to neck on four sides.

At the outer edge of the base mat, cross right spokes over the left and weave pairing rows around them as if they were one. Arrange pairs to lie equally spaced, two each on front and back and one near each corner.

The bottle has a rectangular outer base of the same size and shape as the base proper. Put in ½ inch of close pairing around this outer base, weaving tightly around the crossed spokes and fitting the weave to the bottle. Finish off by running the weavers back into the weave invisibly with the bag needle used head first to avoid splitting.

Pull up on all spokes to tighten them.

* With the right spoke of a corner pair and the left spoke of the pair on its right, make a square knot 1 inch up on the spokes.* A square knot is necessary for proper angle of mesh strands. A thin, strong cardboard 1 inch wide over which

these knots may be tied will help to gauge measurement.

Repeat between * to make the first row of firm square knots at an even distance around the bottle from the base and spaced equally, three knots each on front and back and one at each side center.

In the next row, also 1 inch up on the spokes, the knots, which are again made with spokes from adjacent pairs, should fall one on each corner and two each on front and back as equally spaced as possible.

In the third row of knots, the length of spokes between knots must be shortened to accommodate the jacket to the rounding of the shoulder which it should follow. This will mean a shorter space between knots.

The last row of knots should draw up the jacket tightly and be tied close around the base of the neck.

Select two of the longest spoke ends, coil them lightly, wrap them around the neck of the bottle, and tie them tightly in a square knot.

Draw each of the other spoke pairs under the stay this tie provides, and weave in the ends neatly and symmetrically to make a small coil finish around the base of the neck under the cap roll made in the bottle. In order to keep the coil from becoming too large, use only enough of the spoke ends to fasten them securely and finish off the coil evenly. Trim off any excess material when the jacket has dried.

The hanger is a round braid made of four strands of ¼-inch material braided in lanyard fashion as given in Chapter 12.

To fasten it on, slip the ends of two of the strands under the neck coil and with the other two ends tie it firmly in place at the side center with a square knot.

Tie the other end on the opposite side to make the loop handle.

When dry, trim off the ends to about ½ inch.

Jacket with Crossed Spokes and Pairing Stay

This form, though different from the vinegar jug in shape and size, illustrates that the placing of the pairing ring determines the pattern. Both jackets are made with lightly coiled cattail leaves and much the same procedure. However, the pairing rings in the jacket for the vinegar jug are located about where the spokes cross, resulting in a series of open triangles, while the pattern rings for the bottle jacket are put in about midway between the crosses and develop a pattern of Xs around the bottle as shown in Plate 5, Fig. 3.

Even though the instructions for the vinegar jug jacket (Fig. 1) could be adapted to the bottle, to avoid confusion, the instructions for this bottle jacket are given except for the preparation of the material and the directions for the pairing weave, both of which will be found in Chapter 12.

The bottle is 8 inches tall and just under 3 inches in diameter. Like the vinegar jug, it is the product of another day and was contributed to the collection by a friend from Alabama. It has a light cream glaze, shading to a light tan at the shoulder. The cattail forms a pleasing complement to the pottery bottle both in color and material.

Leaves for spokes, of which there must be eight, should be 3 ½ to 4 feet long after the underwater part of the leaf has been trimmed off. Select for constant width and thickness of leaf. Shorter leaves may be used for the pairing but the standard for width and thickness should be rigid.

Twist the eight spoke leaves individually but lightly and weave into each other at the centers to form a square. This will result in sixteen ends for spokes, four on each side. Clamp three of the corners temporarily with spring clothespins and loop a lightly coiled weaver strand, off center for the sake of splicing, around one of the spokes at the fourth corner and begin the pairing weave for the mat to cover the bottom of the bottle. The work is planned to move from left to right, so the base mat cannot be turned over, and any irregularities should be on the upper or what will finally be the inner side of the mat. Make a firm weave, pulling both spokes and weavers in at the corners of the woven square to round them.

When the diameter of the mat reaches that of the bottle base and the weave reaches the spoke at which it was begun, put in another row of pairing close beside the last but in this row, cross the spokes one over the one next to it, and weave them as one. This will help equalize weave and spoke spacing around the outer base, give angle direction to the spokes, and set a pattern of eight spoke crosses.

Hold the weave temporarily with a spring clothespin while the work is examined for spacing of spokes and tightness of weave. It makes no difference whether spokes are lapped left over right or right over left but for both regularity of weave and ease in tightening spokes, one or the other must be chosen and then adhered to throughout.

Set the bottle on the mat and tie them together firmly from neck, over base mat, and back to neck on four sides. Be sure that the mat is the right diameter, the spokes properly spaced and tight.

Put in two other rows of pairing very tightly around the outer base of the bottle, weaving around the angled spokes in turn and ending the weave at the spoke around which it was begun, so coming full circle.

Thread one of the weavers into a bag needle and push it head first, to avoid splitting weaver or spoke, down beside the spoke and out on the outer side of the mat. Dispose of the other weaver end similarly by threading it into the weave by the side of the adjacent spoke. When both are properly adjusted, they may be trimmed off closely as there is no strain on them.

The weave is now ready for the first independent pairing row. The circumference of the bottle is such that a splice will not be necessary unless the cattails are unduly short; the best part of the leaf can be used to advantage. Coil a weaver lightly, then fold it in the center around a spoke about 1 inch above the last row of base pairing or about the same distance above the natural cross of spokes as the base pairing is below it. Make a cross of the weavers around each spoke and one between them, fixing both spoke and pairing row in place.

When the ring reaches the point of beginning, complete the ring by slipping the end of one weaver through the loop around the first spoke, pull tight, and weave back to the left over itself for a couple of inches, fitting it neatly into the coil. Hold in place by slipping the end under a spoke. Weave the other strand to the right in a similar manner. Ends may be shortened but close trimming should be left until the jacket has been finished and allowed to dry.

By the time the position for the fifth ring of weave has been reached, the bottle has narrowed sharply. Do not cross the spokes above this last independent row which will be placed midway between the row at the point of the shoulder and the first of four continuous rows which form the collar. This will make the two last openings between rows slightly wider.

Loop a lightly coiled weaver strand around two adjacent spokes and begin the collar by weaving around the spokes from pair to pair without crossing the weavers between them. This will make a slightly long weaver coil on the first row which pulls in on succeeding rows if the weave is made tightly. The fourth row of continuous weave should end just below the slight protrusion of the bottle about ¾ inch below the mouth. Dispose of the weaver ends by threading them, one at a time, in a bag needle and running them down into the weave invisibly. Tighten ends to make a smooth upper edge.

Make a final examination of the work, giving any necessary adjustments. Spokes may still be tightened by pulling up on them if they have become slack. Dampen the spoke ends thoroughly before finishing off.

The decorative finish around the neck is made as follows:

* Twist the left spoke of a pair tightly, bring it in front of the right spoke, behind the left spoke of the pair on its right, and down into the weave between the spoke pair by use of the bag needle head first.

Twist the right spoke of the pair tightly, bring it in front of the left spoke of the pair on the right, behind the right spoke and down into the weave by the right side of the right spoke of the pair. *

Repeat between * to weave in all spoke ends. This will make a small, neat scallop edge and dispose of the ends of the spokes. These need to be woven in only to the lower edge of the pairing which makes up the collar. Pull down evenly to make a symmetrical finish and trim off at the edge of the pairing when dry.

A stopper, which is optional, is made of well-moistened short lengths of leaves tied together firmly with a length of twisted cattail at the center, folded to a U shape, tied again about 1 inch above the fold with the ends of the tie sewn through the bundle and up to be trimmed with it to about 3 inches. Shred the ends coarsely with the bag needle.

For the 6-inch cord, twist an even-width leaf of about 18 inches tightly from both ends, pass one end under a pair of spokes, and coil the strand on itself to make a two-strand smooth cord.

Tie the ends of the cord around the stopper where it shows above the top of the bottle and use the bag needle to thread them into it.

Cattail-rush Jacket in Pairing Weave for a Large Straight Bottle

The shadings in the cattail rush show up well in the jacket illustrated in Plate 5, Fig. 4, which is made over a form 13 inches high, 7½ inches in diameter at the shoulder, with a 5-inch-diameter base. These dimensions are not important except as they form a base from which any necessary adjustments to the form in hand may be calculated.

Spokes should be 44 inches long as they must pass entirely around the form with 2 inches extra on each end to allow for ease in finishing. These spokes were of jewel rush cut at positions on the rod to yield a similar size. Any spoke material which is of constant size for the desired length may be used.

Long, narrow cattail leaves were selected for the weavers, with the underwater growth and the clasping edges at the base trimmed off. This part of the leaf is very tough and pliable, but it presents a color and texture at variance with the soft grey green of the other part of the leaf. A use for these parts of the leaves is illustrated in Plate 5, Fig. 2.

Cattail cures readily and is easy to handle but neither it nor the spoke material should be bent sharply while green or dry. Such treatment will almost certainly make a crease and result in a future break. Instructions for harvesting, curing, and moistening cattail rush will be found in Chapter 12, as will also the pairing weave directions.

A very close weave was desired for this jacket, so twenty-eight spoke rods for the fifty-six spokes were selected and moistened, along with the long, thin cattail leaves to be used as the weavers.

To make the base of the jacket, place six rods over six rods at a right angle with centers even, and hold them with spring clothespins.

Place four rods, centered, under this cross on a diagonal and hold in place with clothespins.

Place four rods on the other angle of intersection, all centers even, and hold with clothespins.

This layout provides forty of the spokes needed. The other sixteen will be inserted farther out in the base weave so as to reduce bulk at the center.

Wipe down and coil a cattail leaf lightly, pass one end around the last placed group of four rods close against the groups of six over six rods and weave. Continue the pairing around the groups, separating the groups of six into groups of two, to make the first row of pairing weave, fashioning a neat, firm center for the jacket base.

Put another row of pairing around the groups to stabilize the rods.

Put in a third row of pairing and separate the groups of four into groups of two rods.

Weave a fourth row of pairing close against the third row.

* In the fifth row, beginning at the left side of a former group of four, weave in the first of the remaining eight spoke rods, one at a time, by bending at the center and inserting one end under the upper half of a pairing in the fourth row. So anchored, the insertion will not be noticeable on the outside of the jacket base. The new pairs will also help fill the wide spaces at these angles. Center the inserted pair and weave.

Weave over the next two pairs of rods.

Insert the second rod in the next angle in the same manner and weave.

Weave over three pairs. *

Repeat between * to insert the remaining spoke rods and to complete the pairing row. The pairing rows should be close together and continuous, with no finished rows until the entire jacket is woven.

Weave a row of pairing around spoke pairs.

Put in another row of pairing while separating the pairs and weaving around individual spokes, placing so as to have them evenly spaced by the time the base mat reaches a diameter of 5 inches.

Put in rows of weave as necessary to make a firm, closely woven mat.

Set the bottle on the mat, centers even, and tie them together from neck through mat and back to neck on four sides with a small, strong string.

Proceed with the pairing, drawing each spoke tightly against the side of the bottle, keeping the spokes vertical and evenly spaced.

At 3 ½ inches from the base, change the routine to make a 3-inch-wide band by including two spokes in each pairing. Correctly put in, this does not change the position of the spokes and forms a variation in the weave which is rather pleasing. Do not change the routine of one cross between spokes. A long, easy, flowing coil will fill the weave easily.

On completion of the 3-inch band which should be measured frequently near the end for accuracy, revert to the pairing around each spoke until the harrowing of the form above the shoulders necessitates the combining of spokes.

Forms will be different, so that no absolute directions for diminishing the spread of the weave by combining the spokes can be given. However, they will likely be widely grouped at first and care should be taken to see that they are symmetrically placed with the view of spacing for further necessary spoke reduction.

After the shoulder is reached, the ties which held the jacket in place will no longer be needed and may be cut out.

All reduction of spokes should be accomplished by the time the base of the neck is reached, and there should be only enough spokes to cover the neck with the necessary weaving spaces between them. Put in a firm, even pairing weave over them to just below the mouth of the bottle.

When the weave makes an even row around the neck, finish it off by running the weavers down under the jacket invisibly. The bag needle used eye first to avoid splitting the weave will make this possible and allow the ends to be brought out so the weavers can be tightened.

Do not trim closely until the spokes have been woven under and the jacket has dried.

If the spoke ends have dried out, they must be moistened by allowing them to lie for an hour or so in a wet towel.

The spokes around the neck may have been combined in pairs and woven as one, in which case the left spokes should be used to make a small scallop finish around the neck, leaving the right spoke of the pair to be trimmed off just under the top of the scallop. To make this scallop, coil a spoke lightly, thread the end in a bag needle or other needle with a big eye, pass it across the front of the spoke at its right, behind the third spoke pair, and down into the weave by use of the needle eye first.

When all the spoke ends have been woven in, pull the scallops down to make a close, even edge. The ends should be brought from under the weave near the base of the neck. They will then be fastened and ready for trimming when the work has dried.

The round braid handle, 17 inches long as finished, is made of four of the longest, slenderest leaves available, used full length. Shorter leaves may be used but a nicer braid can be made if splicing is avoided. Directions for this braid are given in the general instructions.

Usually a round braid at least 1/8 inch in

diameter requires a core, but cattail is so pliable and packs so well that in a braid of this size if provides its own padding.

Pass the butt ends of the four leaves through the bottle handle to extend three inches on the right.

Coil these ends lightly, match them with their leaves on the other side of the handle, and braid tightly together into the round braid to fasten the end around the bottle handle.

Braid tightly until all the length of the leaves is used. The taper of the leaves will taper the braid symmetrically to a point.

Pass this end through the handle and bring the end back to be wrapped four or more times around itself for a decorative finish.

Using the large-eyed needle, slip the end back under these wrappings and pull tightly on it while stroking the wrappings to give them firmness.

If other fastening seems desirable, a narrow strand of well-moistened coiled cattail in a raffia needle will stitch it invisibly.

The base may need flattening, and a small, hard disc centered on the base under a press will be helpful for this.

CHAPTER FIVE

JACKETS OF ROUND REED AND CANE

THE JACKETS in this chapter are made with reed spokes and a cane binder, with the exception of that for the demijohn which is made altogether of round reed. Round reed is specified because reed also comes flat, and in various widths, both styles being seen often in basketry.

Cane for the binder is also a commercial product seen frequently in basketry procedures but used most often in seating chairs. These all are products of the Orient, where they have long been items of commercial importance.

Reed, a fairly porous and pliable fiber, does not require a prolonged moistening period unless it has to be manipulated in close turns. Cane, on the contrary, is very thin, hard-surfaced, and brittle, but if it is kept straight, it is difficult to break except at a flaw such as a bad joint or a crease. It requires careful handling, for any crease when it is dry will almost certainly cause a break. The strand should be examined for defects and any found should be cut out. The resulting shorter pieces of cane are useful, especially in bottle jackets where the shorter lengths are easier to handle. Both reed and cane should be coiled individually and the coil fastened lightly before it is moistened for ease in handling. This will permit the removal of strands as needed without disturbing others ready for use.

Cane requires a longer moistening period than reed, so they should be soaked at different times. Cane does best if, after a short period of soaking in cold water, it is allowed to lie in a wet towel overnight. Fifteen minutes of soaking in cold water with a couple of hours in a wet towel will be sufficient for the reed.

Demijohn Jacket in Spaced Over-and-under Weave

Jackets around such bottles as this one (Plate 6, Fig. 1) were put on originally for protection, as most of them break easily. This one, however, is quite sturdy and has survived the hazards of commerce and time.

Usually the jacket was a close, over-one, under-one weave of willow, small round reed, or flat reed, producing a covering both functional and handsome.

The green glass of the bottle contrasts attractively with the natural reed of the jacket. The form is slightly wider than tall, measuring 12 by 14 inches, and a 6-inch neck gives it very pleasing proportions.

Experience in jacketing other bottles is recommended before attempting this one. The weave is entirely of No. 4 round reed. As the strands are both spokes and weavers, they must be long enough to reach entirely around the bottle with allowance for decorative angle and finishing border. A length of 56 inches is recommended for a bottle this size and though some strands will be too long, border weaving length will be ample.

The weave is made up of six groups, each containing three pairs of strands, or a total of thirty-six strand ends for the entire jacket which is begun at the top rather than, as usual, at the base.

The reed needs to be moistened only lightly at this point as there is little strain on the strands until the weaving of the base.

On a flat surface, weave three double strands into three double strands spaced about ½ inch apart. Do not weave the strands as singles or one at a time, but weave the pairs as one to help secure the same tension. Make these strands at a sharp angle so the two inside pairs will lie at a near right angle in the middle of the strands. Secure with spring clothespins.

Make two other identical groupings and hold them with clothespins.

Weave left and right ends of two of these groups into each other so as to form a cluster of four diamonds opposite the cluster made by the weave of three pairs into three pairs. This will form one half of the hexagon neck opening, with inner strand pairs spaced at two inches to outline the opening.

Arrange the weaver spoke strands symmetrically and allow them to ray out in pattern as the shape of the bottle and the progress of the work demand, weaving the double strands alternately over or under when two pairs of strands meet. There are certain geometric forms which will appear in the weave and may be used to check correctness and precision of the weave as, for instance, the diamond made up of four small diamonds below each point of the hexagon, or the shield outlined on the center front.

The double strands which make up the groups of three should expand naturally as much as the shape of the bottle will allow, spaced as regularly as possible. Design emphasis demands that spacing between groups be wider than that within the groups. After the jacket has passed beyond the weave around the neck, many of the spaces will become as wide or wider than the entire group.

When the weave has reached to 4½ inches above the base, check for correctness of weave, tightness, and placing of strand pairs or other irregularities and pin a muslin strip tightly around the bottle above and below the point of greatest width. This will help hold the weave in position, and will hold jacket in place so the bottle may be turned over and placed on the peg holder.

Draw the weave in symmetrically to fit the bottle, fixing it in place temporarily with a tight pairing weave of raffia or soft, strong string. Check for spacing of strands and an equal distance from the base.

Have ready a well-dampened strand of reed at least two and a half times the circumference of the

bottle at this point and put in one row of pairing very tightly against the temporary row, pulling it in to make a snug fit. Fasten off the weaver strand ends by weaving them back over themselves as inconspicuously as possible. Leave short ends to be trimmed off when dry.

The experienced worker will have learned that usually there is no crossing of weaver strands between spokes in the pairing weave. However, the spoke pairs in this jacket are at some distance from each other so that the usual weave is not possible, nor will there be any certain interval between spokes, so that the number of crossings may be specified. When the work has been thoroughly checked and a satisfactory spacing of strand pairs has been obtained, fix them in place by making the pairing fit the space with the requisite number of turns, weaving in the spokes as they are reached.

Set the bottle upright, and remove the muslin band and temporary pairing. Inspect the work thoroughly, particularly the placing of spokes and the tautness of strand pairs. Unless these are of the same tension they will open up rather than lie close together. A pull on the ends separately, then

PLATE 6. Reed and Cane Jackets:
Open weave demijohn jacket or small round reed (fig. 1); diagonal weave of reed and cane made with a slip knot (fig. 2); reed and cane on a rectangular bottle with collar (fig. 3); reed-and-cane jacket for a square bottle with cane overlay (fig. 4).

together, will tighten them if a slight correction in placing does not bring them into line.

Have ready several full-length strands of well-dampened reed.

With the bottle on the peg holder, loop one of these strands off center around the crossing of two spokes about ¾ inch below the single row of pairing and put in the first of five rows of pairing weave. The first row must establish a tight fit with pairs lying at a symmetrical angle. As the pairs are woven together where they cross, there will be no difficulty with a slipping weave as is usually the case with a diminishing form. Single pairs will naturally hold the position. In the next four rows, bring the spoke pairs down gradually so they will lie almost vertical to the base and about 1 inch apart. Fasten the weaver ends by slipping them back, one at a time, into the weave. Cut off closely when the jacket has dried. The last of these rows should be 2¾ inches above the base of the bottle.

Loop a strand of reed, off center, around a pair of spokes 2 inches above the base and put in the first row of a 2-inch base weave, leaving a ⅝-inch open space all around the bottle. Hold the weave with a clothespin and measure for the accuracy of this space. At this point, a little shifting is possible.

Continue the weave until it is even with the bottle base and forms a tight even band all around the bottle. Fasten off the weavers by running them back into the weave invisibly.

Pull up on all spoke pairs to take up any slack. This is the point beyond which no tightening is possible.

Moisten the ends of the weaver spokes by placing a heavy, well-dampened towel over them while the bottle is still on the peg holder. This will take some time but at this point dampening by setting the bottle in water will likely damage some of the strand ends.

When the strands are pliable, pass each pair of spokes behind the adjacent pair on the right and to the front. Slip the last pair, from the back, through the loop made by the first pair, thus making a row of small tight scallops around the base of the bottle close to the 2-inch base weave. Examine for a close, regular scallop.

Bring a pair of spokes in front of the first and second pairs of spokes on its right under the first two pairs of protruding ends, and back toward the center of the bottle through the scallop.

With all spokes woven in this routine, the finish will present the appearance of a flat braid.

The spoke ends will lie at a sharp angle around the base and should be cut to a length of about 2 inches. They can then be pushed under the edge of the base for a smooth finish, forming a protective base for the bottle.

Jacket of Reed and Cane Using the Slip-top Knot

The warm brown of the pottery bottle in Plate 6, Fig. 2, makes a satisfying background for the cane and reed jacket as it seems to intensify the tan of both mediums, and the straight bottle furnishes a perfect form for the diagonal weave, serving also to illustrate the useful slip-top knot with which the intersections of cane and reed strands are tied together.

The weave may be made successfully on bottles with curves, but it is slightly more difficult because of the necessity to expand or diminish the length of the weaver between the spokes. A good design for a straight bottle has the same spacing between both spokes and weaver strands. In the layout given for this bottle, if the length of the weaver between the knots in any row around the bottle is the same, the other dimension will adjust to it.

It is almost impossible to keep the spokes in this weave to a true vertical and, as the spoke angle adds interest to the design, no effort was made to do so.

The bottle is 9 inches high by 2¼ inches in diameter. Sixteen strands or eight pairs of No. 1 reed 24 inches long were allowed. This apparently excessive length is needed to provide for angle and for ends long enough to make the collar with ease.

Fine-fine cane was used for the eight strands, one for each pair of spokes. The cane strands do not pass across the bottom of the bottle but are inserted, and should be two and a half times the height of the bottle.

Allow extra reed for the base weave and two short lengths of cane for the pairing weave around the neck.

Soak all reed and cane strands in cold water for fifteen minutes and wrap in a wet towel to make them pliable.

Weave the eight strands into each other at their centers in pairs.

Weave around the square so made, tightly, in over-one, under-one weave with a reed strand the end of which has been allowed to extend 5 inches on the left as the extra spoke needed for a continuous weave.

Weave the pairs as one, making a mat to cover the bottom of the bottle and extend up the side for about ½ inch. Fit the bottle tightly in, equalizing the space between the spoke pairs.

When the weave has reached an even width around the bottle base, about eight closely woven rows, run the end of the weaver and the extra spoke back into the weave invisibly.

Insert the end of a cane weaver strand into the base weave on the right side of a spoke pair and anchor it securely. Place all eight cane weavers firmly in the base in this manner.

The weave is a series of slip-top knots made with

the cane around each pair of spokes at measured intervals. The width between reed spoke pairs and cane strands must be regular if a workmanlike product is to result. A gauge of thin, strong cardboard will be helpful in measuring the spacing. In this bottle, the spacing was the same, so a measure ¾ inch wide served for both.

The slip-top knots carrying the weave forward may be made on the bottle, but the job is easier done by measure with frequent fittings. Complete one row of knots before beginning on another. Six rows of knots for this bottle brought the weave to just below the shoulder. From this point, the jacket must be made with the bottle in it.

Diminish the space between spokes slightly and put in a row of knots above the shoulder equidistant with those below it.

Put in another row of knots tightly at about ½ inch space. These should be about at the base of the neck and will draw the spokes in to fit the neck of the bottle, and completing a row of diamonds around the neck.

The slip-top instructions will be found in Chapter 12.

With a well-moistened cane strand, put in two rows of pairing weave tightly around the neck ½ inch farther up, straightening the reed spoke pairs to the vertical while doing so, placing the cane weavers around the spoke pairs to catch them firmly into the weave. Make only the regular one cross of weavers between spoke pairs. This will give a tighter weave, yet will allow any slight shifting of spokes needed for adjustments. Fasten off the pairing at the end of the second row by fitting the ends back closely into the weave with short ends to be trimmed off when dry.

Put in two other rows of pairing weave 1 inch below the mouth of the bottle and fasten the ends securely.

The cane weaver strands which have been carried along under the spoke pairs may be cut off above the last pairing row at this point.

Moisten the ends of the spoke pairs so as to have them very pliable.

In a continuous movement, make a loop from the back with the ends of a spoke pair around the pair next to it on the right, bring it back to the left behind its own base and the spoke pair on the left and out to lie pointing left.

Weave all spoke pairs in this manner, slipping the ends of the eighth pair, from the back, through the loop made by the first pair.

Pull the loop ends down at a right angle to the bottle to make a small, even scallop which fits tightly around the neck and leaves the ends lying so they may be cut off close after the work has dried. This may be easily and safely done with a razor blade if an emery board is held under it and the reed end to protect the jacket during the trimming.

Rectangular Bottle with Reed and Cane Jacket

The bottle illustrated in Plate 6, Fig. 3, is only 8 inches high, 3¾ inches wide, and 2¾ inches thick, but it requires a 32-inch length for the twenty-eight strands of No. 3 reed. Size fine or fine-fine cane should be used.

The method of fastening the base strands together, not given previously, should be detailed here. The weave consists of a series of tight alternate wrappings with two strands around the perimeter of the base over crossed strands to fasten the front to back spokes and the side spokes together with a row of very firm weave which looks like a braid on both upper and lower sides.

The spokes pass entirely around the bottle and form a base made up of two layers of spokes which lie one over the other at right angles, held together with the single row of weave around the base.

About 9 inches of cane per woven inch is required. The bottle shown is 13 inches around, and a strand length of at least 117 inches plus a few inches extra for weave takeup and ease in finishing should be allowed for the double weave, if splicing is to be avoided.

Moisten the cane by soaking for fifteen minutes in cold water and wrapping in a wet towel to lie until pliable. Two other strands about two and a half times the circumference of the bottle and a long strand for the neck weave should be at hand but do not need to be moistened until later. Coil all cane lightly and fasten the end in place for ease in handling and to avoid damage to the cane. Dip the reed strands in cold water and lay them under a wet towel. These are not given any great manipulation until nearer the jacket finish so they do not need deep moistening.

Centering of the spoke strands makes little difference at this point; the reed will slip readily under the cane weft and the strands can be made equal after the first side of the base is completed.

Lay the center of the cane base weaver strand diagonally under two spokes crossed at a right angle, vertical over horizontal.

Bring the left end of the cane weaver up, back over the crossed reed spoke strands diagonally, and down under the lower or base spoke, where it lies pointing left and under side up.

Bring the right end of the cane weaver up, over the crossed spoke strands on the other diagonal so as to make an X, and down under the base spoke where it lies underside up and pointing right. This completes a corner which, if well made, will help firm the weave and form a true right angle. Centering the cane weaver at this first corner will give a finished beginning and two equal-length strands with which to weave. Each weaver is fastened down on the underside by the subsequent movement of the opposite weaver. For best results, the weave must be made tight and held in place. The routine is as follows:

*Lay the second front-to-back spoke over the base spoke strand.

Bring the left weaver up in front of the new spoke strand to cross to the right over both strands diagonally and down under the base spoke to lie underside up, pointing left.

Bring the right weaver up in front of the new spoke strand to cross to the left diagonally over the crossed spokes and the first diagonal and down close under the base spoke to lie underside up, pointing right.*

Repeat between * until there are sixteen front spokes fastened on the base spoke, or first side spoke, securing the corner with an extra diagonal. If the weave is made tight, the width between strands will adjust itself evenly. If the base is just slightly large or small for a good fit, the weave may be adjusted by slipping it backward or forward on the base spoke strand. Equalize the spoke strands, allowing for the length of the back strand to cover the base of the bottle.

Turn the corner by passing the outer or right weaver over the cross as usual but bring it up between the horizontal and the vertical strands on the outer side. Pass it down between them on the diagonal which will place the weaver underside up parallel to the vertical strands. The other weaver will lie under the corner cross and out to be centered between the horizontal and vertical spoke cross. Pull both strands very tight.

Lay a side spoke strand parallel to the base strand under the sixteenth front-to-back spoke strand to which the side spoke strands will be fastened.

Tighten the strands and proceed with the weave as given between *, being sure the weavers cross properly on the back to hold down the weave and preserve the pattern. All spoke strands will have been used when the twelfth side spoke is attached to the base strand.

Equalize the side spoke strands, turn the corner tightly, and try the weave on the bottle base for size. Make any necessary adjustments.

Weave the twelfth spoke to the back of the fifteenth spoke strand using directions given between * and follow with the other front-to-back strands.

Turn the corner and weave the other side strands to the first front-to-back spoke strand. Make a secure corner and adjust to a true rectangular shape slightly smaller than the bottle base. Fasten weaver ends firmly.

Dampen the weave well and tie the base securely to the bottle with soft string from neck through base and back to neck on all sides. Properly plac-

ed, these will not interfere with the weave and may be cut out when no longer needed.

With a well-dampened cane strand, put in a single row of pairing weave about ½ inch up on the side of the bottle or about 1 inch above the row of base weave, fitting the spokes in at regular intervals tightly on all sides and at the corners. Fasten the weaver ends by running them back into the pairing invisibly.

The weave is planned to have two spokes outline the bottle on each side of the corners, or four to each corner. This will leave twelve spokes each on front and back and eight on each side with which to develop the pattern.

Begin at the center of either front or back and cross spoke 7 (from the outer side), over spoke 6. Follow alternately with the other ten spokes in a loose braiding technique to make a pattern of diamonds. As the first strand was lapped left over right, this sequence must be maintained.

When the weave is completed on one section, hold it in place with a heavy rubber band placed near the top of the bottle below the shoulder.

Repeat on the reverse side and slip the ends under the rubber band.

As was done on the front and back weaves, cross the two middle side spokes and work out and up in an over-one, under-one procedure to make a diamond-spaced design. Observe the sequence of left over right for the sides also (the sequence could just as well have been planned to lap right over left).

Pull up sharply on all spoke ends and adjust to patterned spacing, making the last row of crosses near the top of the bottle panels.

The four spokes at each corner, meant to outline the corners, should be straight and tight with just a little taper so that the pairs will touch at the shoulders and accent the shape of the bottle. The irregularly shaped spaces between the corner spokes and the other parts of the jacket design are planned to accent the pattern. Incidentally, this is only one of many possibilities for this layout and there is no reason why the worker should not develop an original pattern.

Put in a single row of pairing weave tightly around the bottle about ¼ inch below the point of the shoulder and fasten the ends securely. This will assist in spacing the spokes which make up the pattern. It is almost impossible to put in a pairing weave which will hold the pairs of spokes tightly in place at the corners. This difficulty is overcome by passing one strand of cane twice over both pairs, once above and once below the pairing weave, then crossing these strands at the center between the pairs to pull them taut. This cane cross between pairs makes a decorative knot, much like a bow knot without the loose ends, and it may either be put in at the same time as the pairing weave, using one of the pairing strands, or be applied afterward.

Wrap the bottle and jacket in a wet towel until the reed is pliable so the ends will bend back under the flange without breaking.

A bottle with a flange at the top such as the one shown is not recommended for a beginner, as it makes the reed spoke ends extremely difficult to handle; however, the directions will apply as well to a straight-necked bottle, though some change may be required in the finishing collar.

To combine the four corner spokes, slip a long strand of cane under the two pairs at the base of the neck, center it, and cross the strands to begin a pairing weave, which should extend to about ¼ inch below the flange. The first row of weave sets the pattern for combining the spokes, which should be four groups of three spokes each on front and back. On the sides, make two groups of three spokes each with a group of two spokes between them. The four spokes on each corner should be combined as directed for the first corner. The strings may now be cut as they are no longer needed.

After the first row of pairing is woven, hold the weave with a spring clothespin and examine for grouping, tightness of pairing weave, and placing of spokes. These should rise straight from the shoulder and be kept vertical for the entire neck weave. Pull up firmly on the spokes to take out any slack.

When the weave has reached the place determined upon, finish it off evenly and fasten the weavers securely, running the ends down into the weave invisibly. Cut off closely only after the work has dried.

Examine to see that spokes have not slipped out of place or loosened.

At the top of the pairing weave, cut off two of the spoke ends of each corner group, one from each side, and trim off one from each group of three spokes.

Bring the ends of a pair of spokes across in front of the pair next to it on the right, behind the third pair, and out where it points downward on an angle. Follow this procedure to make a double scallop around the neck under the flange.

Pull the strands down to make an even, close-fitting, double-strand scallop.

Dampen again, for reed does not turn at a sharp angle safely when it is dry, and slip the ends of a pair of spokes over the base of the pair on its right and back under the weave parallel to itself.

Bring the ends out between the flange and the edge of the woven collar for trimming off with a razor blade when it has dried. An emery board or a small piece of cardboard should be held over the weave to protect it from the razor blade.

Jacket of Reed and Cane for a Square-based Bottle

The shape of the bottle illustrated in Plate 6, Fig. 4 seemed to indicate lines, lines at whatever angle, which produced surprising results.

The jacket requires twenty-four strands of No. 4 reed 36 inches long for, though the bottle is only 9 inches tall, it has a 4-inch-square base and the finish at the neck needs 6 inches for developing with ease. At least four full-length strands of fine-fine cane or the equivalent in shorter lengths should be allowed for the pairing weave, which holds the reed spoke strands together, and the angled bands.

Directions for the pairing weave and for the slip-top knots which fasten the strands on the corners will be found in Chapter 12.

For an alternate base weave, see the instructions for the base of the rectangular bottle jacket (Plate 6, Fig. 3).

Moisten cane and reed by soaking a short time in cold water and wrapping in a wet towel to make it pliable. Do not bend the cane strands sharply, for any crease when dry will almost certainly cause a break. Coil the strands individually over the hand and fasten lightly for ease in handling.

Make a square base for the jacket by fastening the twenty-four strands of reed together with a pairing weave around the four sides of a square made by placing the reed spoke strands twelve over twelve, a pair at a time, at a right angle. If the pairing weave is put in symmetrically, the spoke strands will fall automatically at a spacing of ¼ inch.

When all twenty-four spoke strands have been paired and the base mat woven to within about ¾ inch of the base size, true up the mat and put in other rows of pairing if necessary to make a jacket base which should be slightly smaller than the bottom of the bottle.

This treatment will leave the spokes exposed in the center of the mat, the size of the opening depending on the distance from the center at which the weave was begun.

At a space of ½ inch, put in a single row of pairing, pulling tight to fit the outer base of the bottle.

Adjust the bottle on the base mat so the corner spokes fall on each side of each corner.

Check for accurate spacing, adjust the row of pairing to ¼ inch above the base of the bottle, tighten the pairing, and fasten the row of pairing by running it back on itself invisibly.

Put in a second row of pairing around the outer base of the bottle ¼ inch higher up. These rows will furnish the base for the angled applied lines so they must be put in very tightly, placed accurately, and fastened securely.

Dampen the work with the bottle in it.

Press the spokes in firmly at the base of the neck and hold them with a very tight rubber band.

Pull up on all spokes to tighten and adjust the spacing so that the spoke pairs will lie accurately.

Put in a row of pairing at the base of the neck very tightly, weaving two spokes as one and being sure that the spokes on each side of the corners are paired. With the rubber band at the base of the neck holding the spokes down, for this row of weave, it will be necessary to use an expedient in putting in the pairing weave, which requires free spoke ends. For the same effect, loop a weaver around a pair of spokes at the very base of the neck below the rubber band and proceed as follows:

Slip the upper weaver under the first pair of spokes on the right and bring it up between the first and second pairs. Pull tight and hold with the left thumb.

Slip the lower weaver over the first pair of spokes on the right, above the upper edge of the first weaver, under the second pair of spokes, and bring up between the second and third pairs.

Repeat this movement alternately with the weavers to make a close, firm row of weave around the neck, draw it together tightly, and fasten it securely.

Pull upward on all spokes separately and in pairs and check again for placing and proper pairing of spokes. Remove the rubber band.

Put in a second row of pairing immediately above the one woven at the base of the neck. Make tight and fasten securely. This row, of course will be woven in the usual manner as the spoke ends are free.

One inch from the top of the bottle, put in a row of pairing over four spokes at a time, giving attention to spoke sequence, as they must lie straight and in order. Follow with two other rows and fasten off securely.

The angled bands should now be applied, as proper spacing can be secured, yet any loosening of spokes incident to the weave can be corrected from the top.

Put in first the middle strand of the three strands of cane which make up the band by fastening the end of cane in the lower of the two pairing rows around the outer base of the bottle between corner spoke pairs. Carry it over all spokes on the side and fasten it on the point of the opposite shoulder with a decorative slip-top knot, being sure that the strand fits the space and that the knot lies exactly on the point of the shoulder. The last movement in the slip-top leaves the strand pointing left and at the proper angle to be carried over the second side panel and fastened between the corner spokes in the opposite lower pairing row. Repeat to point of beginning. This first strand sets the pattern for the other two strands of the decorative band which

should be applied ¼ inch on each side of it. The end of the upper strand should be fastened in the upper pairing row and that of the lower strand between the corner spoke and the one next to it on each side of a corner. Observe the same spacing of the slip-top knots at the shoulders.

In order to make the pattern of crossed strands on the side panels, it will be necessary to fasten strands at alternate lower corners, securing the ends with as little bulk as possible. The base fastenings are not meant to be decorative, only neat and secure.

Prepare for the uniform sequence of weave for the square in the applied band in the center of each panel by establishing the usual over-one, under-one weave, which, though not vital here, does give nice detail and a note of precise workmanship.

Give a final inspection and tighten all spokes by pulling up on them.

Reed does not make close turns safely unless it is well moistened. Wrap the upper neck and ends of spokes in a wet towel and allow to lie until pliable.

To avoid unnecessary bulk, cut the two center spokes in each group of four, one from each original pair, to about ½ inch from the top of the bottle.

Bring the two remaining spokes of a group smoothly in front of a pair of spokes on the right, behind the third pair, and out between the third and fourth pairs to point downward. Follow this procedure with all pairs of spoke ends and pull them down evenly to make a two-spoke scallop just below the top of the bottle.

Bring the two remaining spokes of a group smoothly in front of a pair of spokes on the right, behind the third pair, and out between the third and fourth pairs to point downward. Follow this procedure with all pairs of spoke ends and pull them down evenly to make a two-spoke scallop just below the top of the bottle.

Make a simulated braid band about ⅝ inch wide around the neck ¼ inch below this scallop, by taking a pair of spoke ends over two pairs on the right, up under two pairs, back over two pairs to the left, and down under themselves and the next two pairs on the left to point downward from under the weave.

Weave all spoke ends in this routine, making an accurate, symmetrical, tight weave. This is an easily handled weave but it must be put in correctly in every detail, pulled down evenly and the ends anchored firmly. At a width of ⅝ inch, it will just about cover the three rows of pairing made around the upper neck, to furnish a finishing base.

Dry thoroughly before trimming off the ends closely. This can best be done with a razor blade if an emery board is held over the weave to protect it from the cutting edge.

CHAPTER SIX

PIGEON-BOX PROCEDURES

THE NAME "pigeon box" in this connection may amuse the reader, but is the traditional one given to the weave by old-time palmetto braiders, who sometimes carried it to the ultimate by working up an entire cabbage palmetto leaf, four or more feet long, into a "pretty," as they called it, measuring about 16 by 9 inches. This was done by using unstripped but halved fronds, still attached to the leaf, to build up forms 1 inch or more at the base and tapering to about 3 inches high tapering to a tip by a series of folds in pattern. The size and number of boxes was dependent on the size of the leaf. As the leaf was used in its natural state, the box size varied according to the place on the leaf, resulting in a heavy center and less bulky boxes at tip and stem end, the whole a very interesting example of pioneer artistry. These were made soon after the fan was harvested and thus cured after being made.

There are three procedures for making the flat pigeon boxes which form the basis for the weaves in this chapter. These remind one of the old four-patch squares for quilts, as they are woven to give the appearance of four small squares of cane, palmetto frond, or other stiff mediums. Given here are procedures numbered 1, 2, and 3 which will be referred to in later instructions by these numbers.

Number 1. This is the traditional pigeon box made with two strands which results in four loose ends. It forms the center of all base weaves and medallions in pigeon-box procedures and so must be mastered. With two strands of the needed lengths, properly moistened for pliability, proceed as follows:

Make a V at the center of the strand with the left end under. With cane, this will mean that the unfinished side is showing on the left.

Lay another strand in the crotch of the V, underside up.

Number strands 1 through 4 beginning at the left.

Lay strand 4 to the left over strands 3 and 2.

Lay strand 2 over strands 4 and 1.

Lay strand 1 over strand 2 and through the crotch of 2 and 3.

Pull all ends to tighten and square the pigeon box.

This weave with its four loose ends, sometimes called a button, is useful as a center weave and is necessary to hold the two extra strands which are a part of all except one of the jackets given in this chapter.

Number 2. This pigeon box is made with two loose ends only and is the procedure used to carry the design forward. Its use is shown particularly well in the jacket for the miniature desk jar in Plate 7. More elaborate designs illustrating its use are the jackets for the camphor bottle and the pitcher in the same plate.

Success with the weave will depend on tightness and accuracy. Gauges over which the strands may be lightly creased at the proper places are necessary and can be made by cutting the right measurements on a strip of thin, stiff cardboard; a small separate gauge for each measure needed is preferable.

Provide an eight-strand end layout made with a regulation pigeon box and two strands inserted on the diagonals into the weave on the under side, with polished side of the cane up.

*Measure two strands and crease them lightly over the gauge for the position of the next pigeon box. When inserted strands are being used, combine each with a regulation pigeon-box strand for stability.

Slip the loop so made in the right strand over the loop made by the under turn of the left strand.

Hold the loops in position and bring the end of the right strand back to the right horizontally, then to the left so that it may be slipped into the left loop. Pull the strand through, leaving a slack loop.

Bring the end of the left strand up between the two strands from the center pigeon box at an angle, over the right strand, and through the loose loop at the upper right corner.

Pull the right strand end to tighten the upper part of the box and pull up on the left strand, which forms the lower right corner, to tighten and square the box.*

This is the procedure used to make the four

short petals which usually surround the center pigeon box at about ½ inch, which does not afford much working space. The second row of boxes is made of strands from adjacent boxes so the box will be easier to make, even though the procedure is the same.

Number 3. This pigeon box, made with one strand, is particularly useful for slings or where decorative knots are needed. The boxes are not quite as true as those of the other two procedures given but they can be made successfully by careful attention to the corners as they are being made. Certainly this pigeon box fills a place uniquely its own.

*Measure the spoke for the position of the upper edge of the box.

Bend the outer end downward at a sharp angle to the right, with the underside of the cane showing.

Bend the strand to the left across the spoke.

Crease the strand downward over the spoke at double its width.

Bring the strand upward on an angle to the right under the spoke and the first movement (that showing the underside of the cane).

Bend the strand to the left and pass the end through the left corner loop to complete the left half of the box.

Bring the strand under the weave from upper left to lower right, up over the lower right corner, and through the upper right loop. This end will be much easier to insert if the tightening of the upper half is deferred until this last weave is completed. The upper half may then be tightened on the underside of the pigeon box and the lower corner by a pull on the upper end of the strand.*

The weave must follow this routine in order to have the outer or finished side of the cane up. For precision and freedom in manipulating the cane, the procedure must be mastered.

Reed and Cane Jacket with Pigeon-box Accent

This jacket makes use of the one-strand pigeon-box procedure to tie the reed spoke pairs together and hold them in place with the cane weaver. At the same time, the one strand of cane required for each pair of spokes helps form a design of diamond-shaped spaces as it moves from spoke pair to spoke pair on an ascending angle. Much of the attractiveness of this jacket, shown in Plate 7, Fig. 1, is due to the very open weave. The bottle is 9½ inches tall by 5 inches at the greatest diameter and is of very dark green glass which makes an excellent foil for the natural color of the jacketing material.

The eight strands of No. 4 reed pass entirely around the bottle and should be 36 inches long to allow for the collar weave. In passing entirely around the bottle, provision is made for eight spoke pairs.

Only four strands of size medium cane are required, but as all the weaving below the collar is done with the cane, the strands should be 60 inches long. Like the spoke strands, they are fastened at their centers on the base and provide the eight weaver strands for the eight spoke pairs.

Dampen both reed and cane by soaking them in cold water for fifteen minutes and wrapping them in a wet towel until they become pliable.

To make the base weave, lay a pair of reed strands, ends even, on a flat surface with their center marked by a spring clothespin.

Lay another pair of strands, centered, across these at a right angle ½ inch from the centers and hold together with a clothespin for two sides of a 1-inch open square.

Center a strand of cane under this crossing of spoke pairs with the finished side of the cane down and, working from the upper side, cross the cane diagonally from corner to corner to make an X on the lower or outer side. Run the cane under, through, and back over a loop to bind the corner tightly and leave the cane strand face down and parallel with the spoke pair. When the cane is moist and this tie is made firmly, it will hold the spoke strands at a true right angle.

At 1 inch from the first corner of spoke pairs, place another pair of reed strands at a right angle to the side pair and bind the corner in the manner described with a centered cane strand. Give attention to the placing of the reed pairs so that both ends of the strands lie either over or under (as the case may be) the pairs at a right angle, or, for example, the side pairs vertical and lying over the upper and lower pairs.

Place the fourth pair of spoke strands in position and tie the last two corners of the square, using a centered cane strand for each.

So far, all the work has been done on the upper or what will be the inner side of the jacket in order to make a smoother base and so that any extra weaving or possible knots will fit into the depression in the bottom of the bottle or at least be on the inner side.

With an extra length of cane which was moistened for the purpose, fill out around the square with a pairing weave to make a close, tight mat rounded to fit the bottle base, catching the cane strands with which the corners were bound so that they will be placed midway between spoke pairs at the edge of the mat. See Chapter 12 for instructions for pairing weave.

The one-strand pigeon-box procedure given at the opening of this chapter is slightly different to this one, in that it is made over itself and this one is made over and woven through the spoke pairs.

*Measure one of the cane strands on the edge of

the mat and crease it lightly at 1½ inches to turn the strand underside up at a sharp angle. This measurement is important as it places the position of the pigeon box.

Bring the strand to the left side of the spoke pair on the right and lay the cane strand under the pair of spokes.

At a slight downward angle, lay the strand back over the spokes below the crease in the cane.

Turn the cane under the spokes and upward at a angle.

Bring the cane up over the spokes again and pass the end through the loop made by the first crease and turn of the cane. These movements have made the left half of the box and the loop at the upper right corner.

Pass the end downward behind the weave and bring it up between the spokes to be woven through the upper right corner loop. Tighten the right corner by a pull on the underside of the strand, which should be left loose until the end is inserted in the right loop, then pull any slack up between the spokes, and finally give a tug on the upper end toward the right and the lower end of the strand toward the left.*

Repeat between * to form pigeon boxes at 1½ inches on the other seven spoke pairs.

Put in two other rows of boxes 2 inches up on the cane. Cane is not meant to withstand the sharp turns and the pressure required to make this weave, so it may sometimes break. If that occurs, remove the pigeon boxes on that strand (each strand being independent of any other) back to the base mat, fasten in a new half strand, and make another set of pigeon boxes.

Up to this point, the jacket may be made more easily off the bottle, as the measurement for placing pigeon boxes is the controlling factor in the weave, and a more exact measure is possible on a flat surface. However, the next two rows of pigeon boxes are measured at 1 inch which will draw the weave in too tightly to admit the insertion of the bottle into the jacket. Also, any tightening of spokes should be done here by pulling up sharply on them at their natural angle. They will slip under the pigeon box, so adjustments are possible.

When the last two rows of pigeon boxes have been put in, the last row at about the base of the neck, pull up on the strands sharply to place them in a vertical position, lay the cane weavers over them, and hold in place with a rubber band. The reed strands will protrude slightly on each side of the cane.

Put in four rows of pairing weave tightly around the upper neck, catching in the cane and spoke strands at even intervals in a close weave, and finish off the pairing about ½ inch below the edge of the bottle. Fasten the pairing strands by running them back into the weave invisibly.

Check the woven band for accurate placement and cut off the weaver strands between the spokes close above the upper edge of the pairing weave.

Finish off the top of the jacket by bringing the left strand of a reed pair in front of the right strand, behind the third strand and out between the third and fourth strands, pulling it down firmly to make a small even scallop.

Repeat between * for a woven edge of sixteen scallops about ⅛ inch below the edge of the bottle. Examine for correctness of weave and symmetry.

For a woven collar just below these scallops, make a band with the spoke ends by bringing a strand under two strands to the right, up and back to the left so that the end may be slipped under itself and the strand next to it on the left.

Repeat between * until all strands ends are woven and pull them down until the width of the band is satisfactory. The one shown is ⅜ inch wide. Strand loops should be pulled down gradually all around rather than trying to reduce them to size with one pull.

Allow the jacket to dry thoroughly before cutting the ends off closely with a razor blade, while protecting the rest of the weave with an emery board or a small piece of stiff cardboard.

Jacket for a Pitcher Using the Pigeon-box Procedure

This jacket (Plate 7, Fig. 3) illustrates the adaptability of this procedure, especially when contrasted with the camphor-bottle jacket and the allover pattern of the jacket for the miniature desk jar.

The general procedure is much the same though the cane is wider and the pattern larger; there are four instead of three medallions with no outlining of them on the upper side. Instead, there is the introduction of chains made up of petal procedures. As in the other examples, pigeon boxes dominate the design and the background is filled in with the connecting strands.

The form is 6¼ inches tall with a 6-inch diameter at the widest and requires sixteen strands of size medium cane 30 inches long for the entire jacket.

If the jacket is to fit, all measurements must be exact, so gauges of ⅝, ¾, and 1 inch should be provided. These may be made of cardboard.

Moisten the cane by soaking it in cold water for fifteen minutes and wrap it in a wet towel to lie until thoroughly pliable.

At the center of two strands, make a pigeon box by directions No. 1.

Insert a third strand under the cross on the back of the weave on a diagonal and center it. Have the finished side of the cane up.

Insert the fourth strand on the other diagonal and center it.

Make four petals ending in pigeon boxes around this center using the ⅝-inch gauge and the No. 2, or two-strand, pigeon-box procedure.

Using the 1-inch gauge, make a pointed petal ending in a pigeon box with right and left strands from adjacent pigeon boxes on the four petal ends.

Make three other pointed petals to complete the medallion and place it under the damp cloth to be kept pliable.

Make three other medallions of the same size, giving careful attention to the tightness of the weave, for this cane is rather stiff and resists manipulation.

Draw the upper right and left strands from pigeon boxes at the petal tips of two medallions together and make a two-strand pigeon box over a ¾-inch gauge.

Draw the lower right and left strands from the same pigeon boxes together to complete the diamond linking the medallions, with a pigeon box at each corner.

Link the four medallions in this manner, fit them around the pitcher at its widest, and link the band in place with the upper pigeon box under the handle between it and the pitcher and the lower pigeon box placed over the base of the handle.

This arrangement of the weave was made to fit the particular pitcher shown. Most other pitchers would likely not lend themselves to such placing of weavers. The aim was to anchor the medallion band firmly. By the time the worker feels equal to the challenge of such a project as this, experience in the necessary adaptations of instructions for other projects should give the skill for any essential changes.

The chains between band and neck are made up of the inner petal formation over a ⅝-inch gauge. Each employs the two strands from the upper pigeon boxes to make a chain of three petal and pigeon-box links above the medallions and four links above the connecting diamonds, all supported by a coil around the narrowest point just below the spout.

To make this coil which is about ¼ inch in diameter, select a chain which has two long ends, coil them lightly, and draw them together very tightly at the point of flare in a square knot, spreading the ends. Use this foundation to hold up the strands of other chains tightly, snugging in the upper pigeon boxes and weaving in the strands, one at a time, to make a firm, decorative coil.

A variation was also made in the weave on the lower side of the medallions. Instead of outlining them with a pigeon box at each point of contact, as in the camphor-bottle jacket, right and left strands from the lower pigeon boxes of the diamonds between the medallions were drawn together in pigeon boxes at the lower edge of the pitcher.

Overlapping the strands from the diamonds, the right and left strands from the medallion points were drawn together to make pigeon boxes between and in line with those made at the edge of the pitcher by the strands from the lower diamond points, to make up a band of eight pigeon boxes at the very edge of the pitcher base.

The design on the base was made by drawing right and left strands from the pigeon boxes on the edge of the pitcher together in pattern tightly and forming a row of eight pigeon boxes about ½ inch from the base edge to give an effect of an eight-pointed star.

The weave was ended in a cluster of four pigeon boxes at the center base, under which the ends were fastened invisibly.

Binding-cane Sling with Single-strand Pigeon Boxes

The bottle over which this sling was made is a clear sea green, contrasting pleasingly with the color of the natural binding cane with which the jacket is made (Plate 7, Fig. 4).

It is 11½ inches high with a 4-inch diameter at its widest, so as each box takes up slightly less than 2 inches, the four strands which make the jacket need to be at least 60 inches long since the strands pass entirely around the bottle with no splicing and make the 12-inch braided handle. Binding cane was preferred for this jacket but the size known as common may also be used. It is lighter in weight and easier to manipulate; the boxes will be smaller, so a slightly shorter strand length will be required.

Two procedures must be learned, the regulation pigeon box given as directions No. 1, and the one-strand pigeon box made by directions No. 3. The pigeon boxes are placed by measurement so insertion of the bottle is not necesssary until after the last two pairs of pigeon boxes are woven.

Coil the strands lightly and fasten. Soak in cold water for a half hour. Wrap in a wet cloth to lie overnight. Binding cane is very stiff and must be made pliable so it will not break and can be handled more easily.

Make a tight pigeon box at the center of two strands using directions No. 1.

Slip strand 3 into the weave on the back diagonally and center.

Insert strand 4 into the weave on the other diagonal and center.

Using the No. 3 or single-strand directions, make pigeon boxes on the eight spoke strands, 1¼ inches from the edge of the center pigeon box. Measure true and draw the weave tight.

Make the second row of pigeon boxes at 1¾ inches.

PLATE 7. Jackets with Pigeon-Box Procedure:
Reed-and-cane jacket with pigeon-box accent (fig. 1); desk jar with pigeon-box procedure (fig. 2); jacket for pitcher (fig. 3); binding cane sling with single-strand pigeon boxes (fig. 4); camphor bottle jacket in the first pigeon-box procedure (fig. 5).

The third box row should be made on the strands 2 inches farther up or 5 inches from the outer edge of the center pigeon box. This will place the row of boxes at the point of greatest diameter. Check to be sure that all pigeon boxes have been made tightly and their positions measured accurately.

Group strands 1 through 4 and hold them with a clothespin.

This leaves strands 5 through 8 grouped on the opposite side.

Measure strand 6 and make a pigeon box 2½ inches from the last box on that strand. Insert the end of strand 5 through the back of the weave on the diagonal so it will lie at its natural angle above the box and pull the pigeon box very tight around it.

Measure strand 7, the other inside strand of the group, and make a pigeon box at 2½ inches, or 7½ inches above the center pigeon box. These pigeon boxes must be evenly spaced if the bottle is to hang properly.

Slip the end of strand 8 diagonally through the underside of the pigeon-box weave on strand 7 so it will lie at its natural angle above the box. Pull the box tight.

Measure for identical spacing on the opposite side of the sling, making boxes on strands 2 and 3 and slipping the ends of strands 1 and 4 through the back of the weave on a diagonal so they will lie at their natural angles above the pigeon boxes.

Measure at 1¼ inches on the same spokes which formed the last pair of pigeon boxes and make the last of the boxes on each side of the sling. Cross the inner spokes one over the other and run them through the weave of the opposite pigeon box on the diagonal. This last pair of boxes will bring the weave to the base of the neck.

Cut a point on a strand of cane the same size as that used for the sling and slip it through the backs of the first row of pigeon boxes twice round to make a stay. Distribute the boxes in a true circle and draw the stay taut.

Use the same procedure to hold the second row in place. Insert the bottle and draw the stay strand tight, being sure that the jacket is centered on the rounded center of the bottle. No fastening of the stay strand will be necessary as the circling of the bottle twice will hold the strand tight and in place. After the jacket has dried, trim the end off closely just past a knot.

Run a stay under the knots at the base of the neck and draw it tight. This is really the strand which holds the jacket on the bottle and is the base of the hanger, so it should be very carefully equalized.

For the worker with no braiding experience beyond three strands, braiding this handle of four strands is just as simple if one remembers that all even-number-strand braids turn one side under, one side over.

Hold the bottle tightly between the knees with the strands even and without creasing them or turning them over. Make a loose, long angled, evenly spaced weave. Stretch the braid frequently to be sure there is no excess of ease or tightness in the strands. Hold with a clothespin.

When braids on each side are about 12 inches long, divide each into pairs and tie with the matching pair in the opposite braid, using single hard knots and being sure both braids are the same length for even hanging. By tying two knots instead of one, considerable bulk is avoided and the ends can be made to point outward at a decorative angle.

CHAPTER SEVEN

WEAVES FOLLOWING THE PATTERNS OF THE BOTTLE

SOME OF the most attractive bottle jackets are those which follow the pattern made in the bottle. These give full scope for the exercise of creative design and sometimes prove to be a test of one's ingenuity.

Finding the appropriate material for the jacket is the first challenge. In some cases, this may be difficult, but in others a material may suggest itself at once. There was never any doubt, for instance, that size fine-fine cane would be the most appropriate and attractive jacketing material for the allover diamond pattern of the bottle shown in Plate 8, Fig. 1, or of craft strip for the lotion-bottle jacket shown in Fig. 4. In fact, the method of weave for the latter is such that few other materials would give such a tight yet sketchy weave.

There are many patterned bottles, so that suitable forms are no rarity. The jacket weaver will do well, however, to confine his choice to bottles with lines — diagonal, horizontal, or vertical — which make more suitable jackets than the more involved patterns and are certainly less difficult to follow.

Sometimes the bottle pattern will allow the elimination of a step not possible on a smooth bottle. For example, there is no cross weave in the jacket for the diamond-patterned bottle, except near the base where the pattern begins and below the shoulder where it ends. These positions were predetermined by slight depressions in the bottle which help hold the two horizontals in place and anchor the diagonal strands, which will remain in place after they have been tightly adjusted. This pattern also illustrates the need to follow and accentuate the original design of the form being jacketed.

Simple weaves give best results. An elaborate weave may detract from the design or obscure the bottle pattern. Lightweight jacketing material will help in avoiding a heavy or clumsy appearance, but serious consideration should be given to the choice of material and the weave most fitting to the form in hand.

The patterns given are merely suggestive and the development of the jacket can be as varied as the bottle. The worker will likely find more pleasure in adapting a weave and material to his own bottle and preference.

Diamond-pattern Jacket for a Diamond-patterned Bottle

The jacketed bottle shown in Plate 8, Fig. 1 is a very simple but effective adaptation of weave to bottle pattern. The pattern in the glass is followed in over-one, under-one weave which is held in place by the depressions in the glass and the tightness of the weave.

The bottle is 11 inches tall and about 3 inches in diameter, with a plain base which extends up the side for 1 inch to a horizontal depression, above which is a row of twelve diamonds. Rows of diamonds extend to the horizontal depression below the shoulder to make up the allover diamond pattern on the sides of the bottle.

Size fine-fine cane was used in the development of this jacket, requiring lengths of 36 inches for the twelve pairs of strands which pass around the base of the bottle and are fastened in the band around the top depression. Material for the stay strand and binder should also be provided.

Coil all cane lightly, fasten, soak fifteen minutes in cold water. Allow to lie in a wet towel until pliable.

Place a strong rubber band around the bottle in the lower depression.

Slip the even ends of two strands under the band and across the bottle, centering base and strands, and under the band on the opposite side.

Adjust to a parallel 1¼ inches between the strands at the center of the bottle. The ends of these two strands will be drawn together to form the lower half of diamonds in the first row centered halfway between them.

Place two other strands across these first strands at a right angle, also 1¼ inches apart, and weave the upper right end over, the upper left end under the first pair of strands.

Reverse the weave to place the lower right strand under and the lower left strand over the first pair of strands. Hold the ends in place with the rubber band (see line drawing A).

If the eye will select and follow a pair of strands

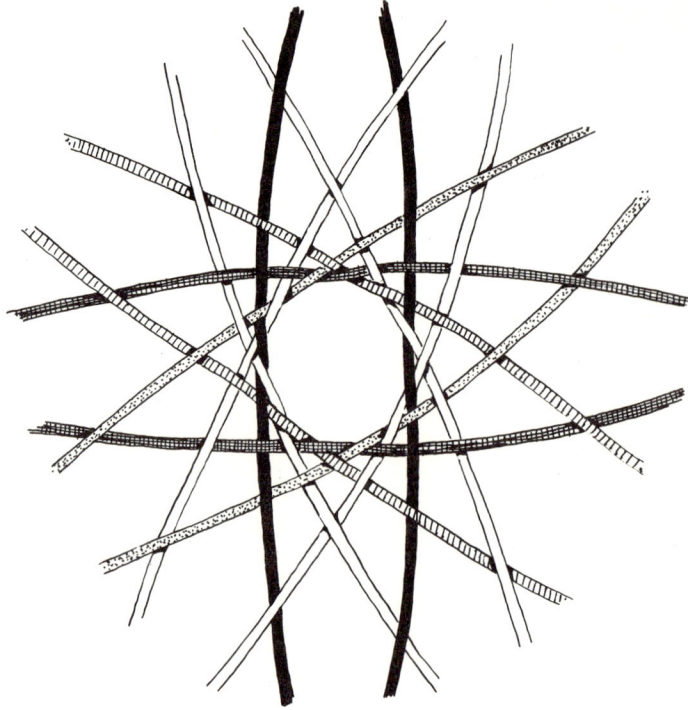

in the drawing to establish their relationship to other pairs, the procedure will be clear.

Follow with the pair of weavers centered on each of the four original strands, weaving each one as the sequence requires, over one, under one or under one, over one, and hold the ends under the rubber band.

These secondary pairs all cross left over right. If the sequence of the weave is properly set up to this point, the counting can begin anywhere on the base and follow in unbroken order but every single step in the weave must be correctly taken.

Adjust and tighten the twenty-four weaver spokes provided by this layout, being sure they are crossed left over right between the base and depression and right over left under the rubber band.

Slip the band up a little and wrap a strand of cane twice tightly around the bottle and weaver spokes at the horizontal depression and fasten it by slipping the end of the strand *down under the stay between the cross of the strands at the point of a diamond, up, over, and back under the stay.*

Repeat until there is a pattern of four precisely placed cane wrappings around the stay betweeen each cross of weaver spokes. Fasten the end by slipping it under the wrappings invisibly.

Once this restraining strand is in and the weaver spokes are tightened by pulling them upward at the proper angle, the body of the weave is very simple, being only the placing of the weaver spokes in the bottle depressions in the proper

order. However, it is vital that the sequence of the weave be perfectly maintained. A good check on this is that the uneven rows of weave lap one way, the even rows lap the other. If any variation crops up, a mistake has been made.

The stay at the top depression is a slightly heavier repetition of that at the base because the ends of the weaver spokes are used to make it, holding them tight and disposing of them.

Check for accurate weaving and tie a soft string tightly around the bottle a little below the upper depression and pull up on each strand at its own angle to tighten it.

Choose one of the longest weaver spoke ends of a pair, wrap it tightly around bottle and weaver spokes at the horizontal depression in the bottle, and slip the end under itself to hold it in place.

Fill in the stay with the strand ends, tightening each weaver spoke as it is woven and cutting out the surplus as necessary to hold down the thickness of the stay after the ends have been fastened. Use one of the long ends for the over-and-under wrappings which will likely need to be cut down to three between each diamond point in order to make it seem slightly less heavy.

Lamp-base Jacket with Outlines Following the Glass Pattern

The challenge presented by the necessity of a continuing weave with a minimum of restraint was the chief attraction of the lamp base shown in Plate 8, Fig. 2.

PLATE 8. Jackets Made to Follow the Patterns on the Glass:
Cane jacket following a diamond pattern (fig. 1); lamp base jacket of binding cane following a leaf pattern (fig. 2); Reed-and-cane jacket for a rectangular pattern (fig. 3); jacket of craft stript for a bath salts bottle, following horizontal lines (fig. 4).

There are six leaves in the pattern so twelve strand ends are required. Select six strands of binding cane 46 inches long, one of which should be at least 12 inches longer than the other five, and weave them together at the center, except for the long strand which should have the extra length on one side. Fasten together at the center with a short piece of cane by looping it once around each of the six pairs of strands. Spread the spokes to accommodate the cord entrance and have ready a dowel or other plug which will hold the strands in place when it is slipped through the center of this weave and into the hole drilled in the bottom of the bottle. Put a heavy rubber band around the neck of the bottle, lap left strand over right at the base (this will help hold them in place), and slip the spoke ends under this band. Have the spoke pairs directly under the tips of the leaves and the pair containing the long strand at one side of the handle. Secure the placing of the spokes from the lower side temporarily with a soft string by an over-and-under movement around each pair just below the tips of the leaves. Such a method of fastening will permit any necessary shifting of the stay. Adjust and tighten the spokes.

Turn a pair of spokes to give an outward angle for the leaf outline. This turn will require two movements in order to turn the strands right side up and will need precision and a tight grip, for these turns will mar the strands so there should be no adjustments at this point.

Have ready a well-moistened strand of medium cane three and one half times the circumference of the bottle and fasten one end temporarily in the rubber band.

Pull the inside strands of two pairs of spokes together about midway up the leaves where the pattern meets, allowing a little ease for the spoke to follow the curve of the leaf, wrap these two spoke strands with the strand of medium cane binder twice, and tie a single knot the second time around by slipping the end of the binder through the loop from the upper side to tie them together firmly. Examine for placing, tightness, and fit. Flatten the spoke on the left side of a leaf and wrap it six times with the binder on an easy angle, passing the last time on under both spokes where they were turned to make the outward angle, wrap the second time, and slip the end of the binder through the loop from the lower side to tie a single knot fastening the spokes together firmly at the tip of a leaf. Flatten the right spoke of the pair and wrap it six times with the binder, or the same distance as for the other half of the leaf, to meet the spoke of the second pair. Repeat between * to outline the lower half of the other five leaves.

For the design to be effective the spokes must be straight and lie closely side by side without slack on the bottom section and one on another where they join at the leaf intersections.

At a step made in the bottle just below the handle, wrap a well-moistened strand of medium cane twice around the neck and tie with a square knot. Medium cane is specified because it will lie closer and tie with less bulk than the binder cane.

Set the outline of a leaf pattern in place, ease or tighten as necessary, and slip the ends of the pair of spokes, one at a time, over and under the cane collar just tied. Follow the same procedure with the other leaf outlines. This will hold the pattern firmly in place. Final adjustments must be made just here.

The two strand intended for the handle should be fastened by a turn over the collar, but, aside from getting in place at the proper angle, they take no part in the weaving of the coil around the collar. This coil has the double purpose of giving a decorative finish to the jacket and disposing of the ends of the spokes. No set instructions can be given but the aim is to make a symmetrical coil with binder strands either angled one way or crossing in pattern, as preferred. As the ends become firmly fastened they may be cut off if not needed to pad or to wrap the coil. Before beginning the coil, adjust the upper end of the medium cane binder to lie invisibly under one of the pair of spokes it ties together, wrapping the end under and over the stay strand tied around the neck of the bottle. Dispose of the other end in the same manner.

For weaving the handle, bring the shorter reserved strand directly to the outside center under the handle and hold it in position by passing the long strand over it, around through the handle, back around, and under the shorter strand, through the handle again, and back over the shorter strand. This gives an over-one, under-one weave at the outside center of the handle and holds the wrapping in place. The weave should be made tight and can be tightened a bit more by pulling downward on the shorter strand at each overpass of the weaver.

When the handle has been covered, lap one strand neatly over the other, pull them around the upper neck of the bottle from opposite sides above the projection made on the bottle for the cap, which will hold them in place, and tie tightly with a square knot. Dispose of the ends by slipping them back under themselves. For such manipulations, the cane must be very pliable, for dry cane breaks. If the work must be left or if the cane dries out while it is being worked, it can always have the pliability renewed by a period in a wet towel. However, overlong soaking is not desirable as long wetting tends to dull the cane and in extreme cases encourages mold.

Craft-strip Jacket Following Horizontal Lines

The inspiration for this jacket (Plate 8, Fig. 4) was in the horizontal rings which make up the bottle design and the black decorative cap which suggested craft strip as a medium for jacketing it. Craft strip also has the advantage of being waterproof. The jacket is not woven but is rather a series of single knots tied at regular intervals to make a very open jacket for the 9-inch tall, 2½-inch-diameter bottle. Three strands of very narrow craft strip 60 inches long are required for the project.

To divide the bottle into thirds accurately, cut a pattern to fit the base and mark it in three equal divisions to fit the positions of the three spoke strands. By placing the base strands on this pattern and keeping the three spokes at a true vertical, the strands will be equally placed around the bottle.

Tie a single knot, loosely, near the end of one of the strands and slip the two other strands through it from opposite sides. Tie a single knot in the end of each, pull then back near the binding knot, and make it tight around the two strands. By distributing the center knot in this manner, it will be less likely to interfere with contact on a level base. Craft strip stretches, so this must be taken into account with all knots, as well as spokes and horizontals, which must be made very tight.

Tie a knot 1¼ inches from the center knot in each spoke strand and hold it open with a large needle or a toothpick. These knots form the first knots in the design and should be so spaced as to barely reach the first horizontal ring. Slip one of the strand ends through two knots and back through its own knot, straighten the strand around the bottle, and pull it very tight, adjusting the knot to fit very closely with the help of the needle. Adjust to position and make the knots tight in the two other spoke strands so this first horizontal will lie in the first depression in the glass. The weave is so shallow at this point that it will be likely to slip off the bottle if it is not restrained, but three lightweight strong strings tied over these first knots and the other ends drawn tightly around the neck of the bottle and tied securely will serve the purpose. These strings may be cut out after the first three horizontals have been put in.

Tie a knot in each of the verticals ⅛ inch above each tie around a horizontal and use the needle to space and tighten. This is the row of spacing knots between each horizontal.

Tie a knot in each vertical ⅛ inch above these and hold the knots open.

The verticals make the horizontals in turn. If you are not sure which strand to use, the length of strand will show clearly. Each strand will have to serve its turn as the height of the bottle and its diameter will require practically all of the length.

Select a vertical strand, run the end through the other open knots and back through its own, straighten the horizontal, and pull it tightly around the bottle. Hold it firmly and tighten the knot so that it lies in the next depression. Adjust

and tighten the other two knots.

Make the second row of spacing knots ⅛ inch above the second horizontal.

Put in the third horizontal of the three-strand group at a distance of ⅛ inch and examine for tightness, spacing, and placing of the vertical spoke strands.

Make four tight single knots spaced at ⅛ inch in each spoke strand to provide the spacing for the horizontal groups. The jacket shown has four groups of three horizontals each. This may be varied but the knots must be very tight and regularly spaced, horizontals properly placed, and the spoke strands on a true vertical if a successful jacket is to be made.

Repeat as directed for the first group of horizontals followed by the spacing group of four knots until the work reaches to about 1 inch below the shoulder of the bottle.

Tie single knots at ⅛ inch to the shoulder and two in each strand after the point of the shoulder has been passed.

Pull up on one strand to make it straight and tight, pass it around the neck of the bottle, and tie it around itself in a firm knot.

Slip the ends of the other two spoke strands under the neck strand, pull up very tightly on these strands at their vertical position, and wrap them one at a time over the first neck strand and its end.

Thread the longest end in a mattress needle and, while holding the strands in the small roll so made tightly, bind them firmly in place with whipstitches around them. Trim off ends as they become fixed to hold the size of the coil to the minimum. Fasten the binder securely by running it under and into the coil.

For a shorter bottle, four strands may be used, requiring only two strands of craft strip. These may be tied together, one around the other at the center of the strip length to be placed on the base. Weave and procedure for this pattern would be the same though the general procedure can have many variations.

Rectangular-mesh Jacket Following the Bottle Pattern

The jacket for this bottle 5¾ inches high with a 2-inch diameter follows a very simple pattern made in only the lower half of the bottle (Plate 8, Fig. 3).

The jacket was developed with size 60 round reed for the twelve spokes or the six strands which pass entirely around the bottle and size fine-fine cane. Five yards of cane should be sufficient for the weavers, or a number of short strands might be used, as most of the pairing rows are put in and finished off singly.

Soak both reed and cane for fifteen minutes in cold water and wrap in a wet towel until pliable.

Cross four of the spokes in pairs at their centers and hold them together with a strand of cane looped around the cross. Put in two rows of pairing around this center, separating the pairs in the second row.

Lay the other strands, all centers matching, across the back of this weave, placing them so as to take advantage of the widest spaces, and fasten them in with the third row of pairing. See Chapter 12 for instructions for pairing.

Continue the pairing weave closely to the 2-inch diameter of the base, equalizing the space between spokes.

Center the jacket base on the bottle and tie them together firmly, neck over base and back to neck on four sides.

Put in two rows of pairing closely around the edge of the mat, pulling the second row in firmly around the bottle to turn the spokes upward around the lower edge of the bottle. This last row should be very near the base of the bottle and hug it tightly.

Put in three more rows of pairing closely and finish off the base weave by running the ends of the pairing back into the base weave using a needle with a large eye.

Pull up on the spokes firmly to make them fall into the vertical indentations in the lower half of the bottle and tie them in place with a soft string around the bottle below the shoulder.

Put in one row of pairing at each of the three horizontal indentations, finishing off each row individually by weaving the ends back into the row. After the jacket has been finished and allowed to dry, cut the ends off closely.

Put in two rows of pairing at the fourth indentation, finishing off at the end of the second row.

The rest of the bottle has no indentation, but one row of pairing about ¼ inch below the shoulder is necessary to hold the spokes to a true vertical with equal spacings.

Immerse the bottle in cold water for a short time and wrap it in a damp towel until the spokes are pliable. They must be pulled in tightly and fastened around the neck. Reed does lend itself to sharp angles readily but not unless it is well dampened.

Twist a well-dampened length of cane lightly, wrap it twice around spokes and bottle neck, and tie it very tightly just below the cap base with a square knot.

The spokes are not sufficiently close together to form the scallop edge which is used in many of the other bottle jackets, so a coil finish was used. The coil should be as small as possible, certainly less than ¼ inch in diameter. In order to hold it to size, the spoke ends must be trimmed out as they become fastened in by the cane binder in equally spaced wrappings.

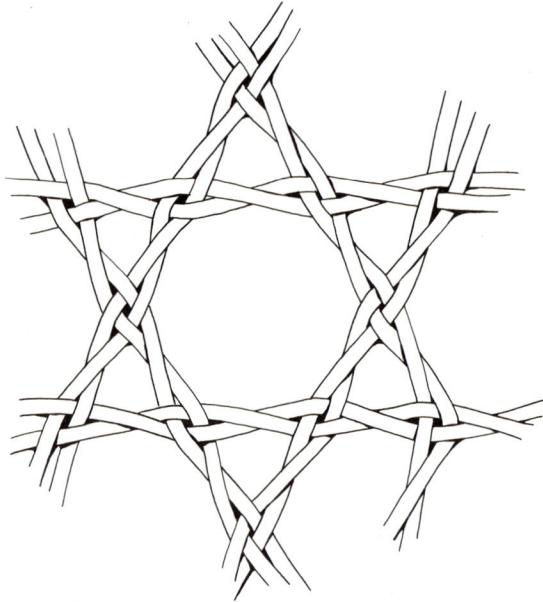

CHAPTER EIGHT

JACKETING WITH THE ORIENTAL WEAVE

AMONG ITS other virtues, the oriental weave is versatile and adaptable to various beginnings, developments, and finishes. It may be a bold mesh or have a modest diamond spacing. It can be made to fit a form with a straight side, a bulged top or bottom, a curved side, or a constricted center and it may be begun or finished at either the top or the bottom. All of these qualities are illustrated in the jackets shown in Plate 9 and in Plate 1, which shows the weave on a lawn-party punch bowl made of a large gourd.

The accompanying drawing shows the weave to be just another development of the over-one, under-one procedure even though in a decidedly devious fashion (line drawing B).

Chair cane is the usual medium used for this procedure but any thin, flat, tough material which can be split to a constant width and made pliable by moistening will give excellent results. A softer fiber may also be used as in the sansevieria-fiber jackets (Plate 9, Fig. 1 and Fig. 5) which also show other developments of jacket bases.

It will be helpful to study the drawing, which shows the arrangement of strands for a smooth, symmetrical jacket base, continuing into groupings of strands for the side of the form and the makeup of the first row of diamond spaces. As the eye becomes familiar with the procedure by selecting a weaver and following it through the drawing, the weave will be seen to progress in a series of four strands woven into a small square with a connecting pair of weavers crossed right over left or left over right, as determined by the arrangement of the weavers in the first square, between each woven group-of-four weave. This crossing of strands between woven squares preserves the sequence and allows the count of over one, under one or under one, over one to begin at any point in the jacket and, if there are no errors, the sequence will be maintained.

The drawing shows a closely woven hexagon with a small woven square at each angle. The triangular points which surround it are made by drawing together adjacent crossed strands into a small woven square at its tip in the proper sequence. This, in large measure, is the content of the weave, not at all involved once it is understood.

Strand count. To determine the number of strands needed for a form, divide its circumference at the widest by the approximate size of the mesh desired which will give the number of groups of four. Multiply this number by four to determine the number of spokes, but divide that number by two because both ends of the strands are used since the strands are centered on the base of the bottle.

Jacket in Oriental Weave for a Vining Plant

The material used for this jacket, sansevieria fiber, is very tough and not at all brittle, splits evenly, and is easy to control. It grows throughout much of Florida and is valued as a decorative plant. Escaped from cultivation, it sometimes grows to 5 or more feet if the conditions are favorable. Its common name of bowstring hemp gives an idea of its character. This jacket (Plate 9, Fig. 1) and the water bottle or vase (Fig. 3) are both made of this material and show the use of multiple fibers for this procedure, in which chair cane is usually used.

The form is sharply tapered 8 inches tall by 9 inches at the largest circumference. Using the formula given for strand count earlier in this chapter, the jacket will require four groups of four strands or sixteen strands to outline its 2¼-inch mesh. As the strands pass entirely around the bottle, this will mean eight double-length strands are needed to make the jacket. These should be 40 inches long to allow for angle, weave takeup, knots, and 10-inch hanger braids.

Another more suitable method for making the base of soft fibers is given for this jacket base. That shown in Drawing A is meant for stiff material.

Split the dry fiber into suitable strands and moisten by soaking it in cold water for a short time and wrapping it in a wet towel until pliable.

The weave is carried out over one, under one. It is unimportant whether right laps over left or left over right but as the pattern is laid, it must be maintained throughout.

The center of the base is a ½-inch open square. Begin the first corner by crossing two strands at a right angle in their centers, left over right, or horizontal over vertical.

Weave in another strand parallel to the first, under one, centering it.

Weave the fourth strand vertically under one, over one, and center it.

This corner weave sets the sequence for the entire weave so must be accurate if the directions are to work out. Hold with a spring clothespin.

*Cross the extensions of the strands on the right side, upper over lower, about ¼ inch from the corner weave.

Weave in a second vertical strand at its center and in sequence.*

Repeat between * for the other two corners. Note that by including the cross between the corner weaver, an over-one, under-one or an under-one, over-one count is possible wherever the count begins. Correct procedure will make this possible throughout the jacket.

Pull the square to size and tighten the corner weaves.

Cross the extension of pairs on a side in sequence and bring them together to weave into a small square at the tip of a ¾-inch triangle above the side of the center square, observing the sequence of the weave.

Repeat between * to make the other three triangles. Hold with spring clothespins and check the work for the sequence of the weave. There should be a square and four triangles outlined by crossed strands between woven corner and triangle tip squares, the whole of such size as to fit easily in the base of the bottle.

Cross adjacent inner-strand pairs left over right, and bring them together in a square woven in sequence to form a triangle with 1-inch sides.

Repeat between * to form three more triangles.

Make a final examination of this base weave for accuracy and symmetry, set the bottle on the base weave, centers even, and draw the strands in to fit tightly around the base of the bottle, adjusting the spacing of the triangle tips if necessary. The small woven squares at the tips of the larger triangle should lie about ¼ inch above the edge.

The second row of outer squares is 1¼ inches farther up on crossed strands from adjacent triangle points. Make the other three woven squares in this row at the same measured distance from the lower row.

Owing to the diminishing size of the bottle, the distance between rows is shortened to just over 1 inch in the third row, decreasing steadily to fit the taper of the bottle. This will make the upper half of the rectangular spaces slightly shorter, but not glaringly so, if the strands are carried forward on a smoothly ascending diagonal.

When the weave reaches the cap roll of the bottle, make a final examination for accuracy of weave, tightness, and angle of strand.

Set aside the inner two pairs of strands of adjacent squares on opposite sides of the bottle neck to make the four-strand hanger braid.

*Bring the two other pairs of strands remaining on one side together in a tightly woven square at the cap roll.

Carry the two outer strands of this square, lightly coiled to reduce width, around the neck above the cap roll and tie them tightly in a square knot on the opposite side above the woven square.*

Repeat between * for the opposite pairs of outer strands and finish all ends with a tightly tied single knot in each end close to the square knot. Cut off ends to within ¼ inch of the knot.

Make two round braid hangers 10 inches long by drawing together neatly the four strands on each side reserved for the purpose. Directions for this braid will be found in Chapter 12. The worker should learn the procedure by practicing with two colors of heavy string before hanger braid is attempted. Regular pattern develops with two colors so that errors show up at once, making the learning process easier.

Tie the braids together in a square knot. Tie single knots in ends.

Oriental Weave Jacket of Split Scrub-Palmetto Stem

This jacket is the result of an experiment to test the practical use of split scrub-palmetto stem. A part of the experiment was the dyeing, though the split stem could very well have been used in its natural state as it has a hard canelike outer finish which polishes nicely. The strands should be split to width and thickness before being dyed. Although the outer surface of the split takes the dye readily, it does not penetrate evenly, and the outer surface is almost impervious to the dye though it did impart a slight warming to the natural color which is still a very light tan.

The jar is 3¼ inches tall with a 10-inch circumference at the base. Using the formula given for strand count earlier in this chapter, twenty-four spokes or twelve strands 16 inches long are needed for the six groups of four with a mesh slightly smaller than 1¼ inches (Plate 9, Fig. 2).

The jar has a glass top so the jacket was finished with a coil around the outside even with the seat made inside to support the top. This disposes of the strand ends and really adds to the attractiveness of the jacket.

The material, if dried after being dyed, will need to be soaked in cold water for a short time and wrapped in a damp cloth.

Using the drawing as a pattern, make a base weave with a 2-inch hexagon opening, the corners of which will fit easily under the base.

Cross the pairs of inner adjacent strands from two of these corners and bring them together in a woven group of four at the tip of a triangle spaced at ¾ inch. The drawing shows the procedure thus far and it remains only to center the jacket base carefully on the jar base and tighten the groups of

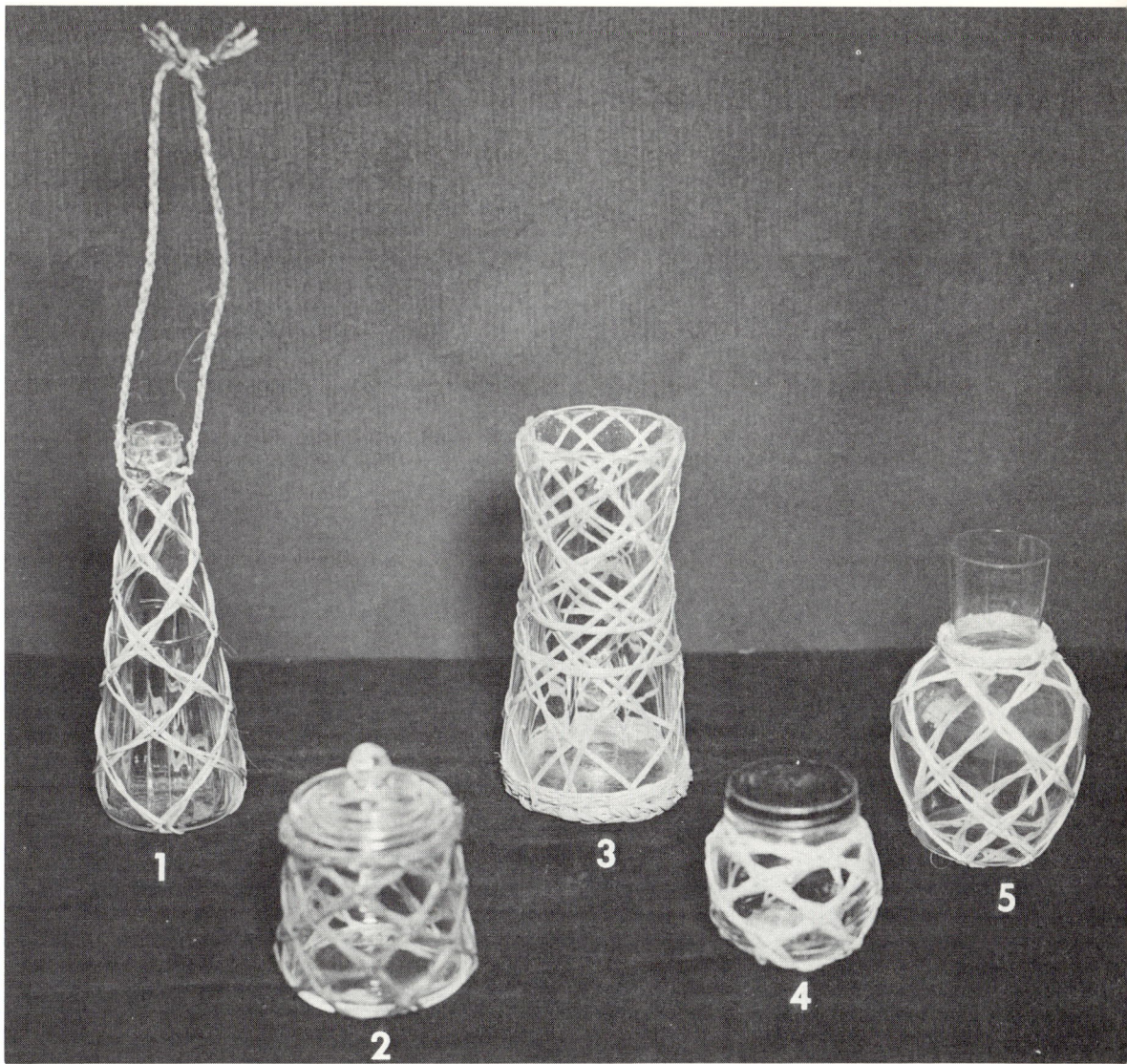

PLATE 9. The Oriental Weave:
Jacket for a tapered bottle with a round braid hanger (fig. 1); covered-jar jacket of split and dyed scrub palmetto stem (fig. 2); adaptation of the oriental weave to a cut-off-bottle vase (fig. 3); jacket for a small jelly jar woven from the top (fig. 4); small water-bottle vase with sansevieria fiber jacket (fig.5).

four at the tip of the triangles to make a row of woven squares around the base just below the turn of base to side.

Make a row of woven squares 1 inch farther up on adjacent crossed strand pairs by drawing them together in a small square at the top of a triangular-shaped mesh, the lower side of which lies on the base below the turn of the side.

Drawing B shows the left strand of the pairs crossing over the right consistently. Once the sequence is established, it must be maintained throughout, although it could just as well have been planned to cross right strand over left.

Space both the second and the third rows of squares on the side at 1 inch up on the crossed strands also. The form has a very slight slope.

The fourth row of squares should be made as usual but the spacing for the row should be just under 1 inch.

The fifth and last row of side squares should be spaced at ¾ inch, so it will be ½ inch from the upper edge of the jar, where the squares will lie about halfway under the decorative coil which finishes the top of the jacket.

Wrap a small, strong twine twice around the jar and strand ends at the upper edge of these squares, pull tight, and tie firmly in a square knot. Examine the work for placing of squares and tightness of weave. Pull up on each strand at its individual angle firmly to take out slack.

Lay the work in a wet towel for the strands to become pliable.

Reserve the two longest strand ends. From the others, make a coil ¼ inch in diameter in a long-angled wrapping of the string, preserving their natural angle. As the strands become fastened, cut off the ends so as to hold down the size of the coil and keep it even.

Wrap one of the reserved strands from left to right around the coil at symmetrical intervals and cross these with the other reserved strand from right to left. Fasten the ends by slipping them into the coil.

Trim off any loose ends closely after the jacket has dried.

Oriental Weave Adapted to an Hourglass Form

This adaptation of the oriental weave shows a decided variation in that it uses a band of parallel outlined diamonds for a contrasting weave around the center through which two weaver strands are run horizontally.

The jacket for this very slightly flared, cut-off bottle (Plate 9, Fig. 3) follows the pattern on an imported vase and, though its shape does not approximate the charm of the vase, it does serve to illustrate the weave.

The form is 7 inches tall with circumferences of 11-½ inches at the base, 10 inches at the top, and 9½ inches at the center.

Calculating for a top spacing of approximately 2 inches, the jacket will require ten strands of size medium cane 32 inches long for the twenty spokes. Though this may seem overlong, the length is needed for a spiral weave, strands which reach from base to top and back again, ending in a built-up base, and for ease in finishing.

The flared top of the vase from which this jacket was adapted required fourteen strands or twenty-eight spokes for proper spacing. This illustrates that almost any adjustment is possible if the basic rule of multiples of four for the spoke count is observed.

All of the other examples given of the oriental weave show weaver strands crossing between each group of four, adjusting to both angle and curve. A good example of the first is the condiment-bottle jacket (Plate 9, Fig. 1) and of the second, the jacket for the jelly pot (Fig. 4). The vase adaptation, however, uses a combination of the usual crossed spokes between groups of four near top and base and parallel weavers between them around the center.

The crossed spokes between groups of four are designed to preserve the sequence of the weave, to take up any slack in the weaver, and to add a decorative note. Though the weave is in some measure self-adjusting, it must be put on accurately and kept tight.

Moisten the cane by soaking it in cold water for fifteen minutes, then wrap it in a wet towel to lie until it becomes pliable. Avoid sharp turns and creases, as they will almost certainly cause a break. Prepare the strands for moistening by coiling them lightly, individually, so they may be taken from the towel as needed without disturbing the other strands. The jacket and form may be immersed as necessary in a cold-water bath to be remoistened.

Place a heavy rubber band around the upper half of the form and slip the left ends of two strands under it on an angle to the left. Center the strands on the top edge of the form.

Slip the right ends of two other strands under the rubber band on an angle to the right and weave the upper strand of the pair over the lower and under the upper strands from the right.

Weave the second or lower strand of the pair from the left under the lower and over the upper strands of the pair from the right. This forms a group-of-four weave of about ½ inch and is one of five such groups around the upper edge of the form.

Center these strands from the left at their natural angles on the edge of the form where, at spaces of about 1 inch along the edge, the lower strand will fall easily over the upper on a long curve, and may be passed under the rubber band.

Lay the upper right strand over the lower right strand along the edge of the form at a long angle to the left and pass the end under the rubber band at its natural angle.

Place another strand under the rubber band on the right above the first pair of strands and weave it over the lower and under the upper original strands from the left, across the edge of the form where it should be centered, to pass over the upper and under the lower original right strands pointing left. Pass the end at a natural angle under the rubber band.

Examine the work carefully for any deviation from the over-one, under-one or the under-one, over-one weave and note that the motif just completed has a widely spaced triangular shape with a small square at its lower tip made up of four strands which pass just over the edge of the form.

The third or inserted strand is a part of the motif on its left and right, forms the upper edge of the small inner triangle below the edge of the form, and shows the interweaving of strands around the top of the vase.

This is the first of five motifs so woven and placed as to anchor the weave over the edge. Doing this successfully requires a well-tightened weave with very pliable weavers. Time taken to run small cord over these first five strands in not less than three colors will be well spent as it will outline the motif so it is readily identifiable and will help train the eye to be selective in a maze of weavers.

Follow with other strands inserted so as to preserve the over-one, under-one or the under-one, over-one weave throughout and form the four other triangular motifs around the top of the vase. Pull the well-dampened weavers down firmly under the rubber band. If the edge strands show a tendency to slip, catch them together lightly from side to side with string woven across the top of the vase. If the weave is put in correctly, the strands will not slip off after they have dried.

The second row of groups of four follows the angle of the strands and places itself naturally at the lower right and left of each motif, the strands between being uncrossed.

To accommodate the jacket to the change in the shape of the vase toward the middle, the strands below this second row of groups of four are crossed in sequence and brought together in a woven group at their tip with a spacing of slightly less than 1 inch. It must be remembered that in counting the weave the over or under of strands between groups of four is of the same importance as that in any other part of the weave.

Make three other rows of groups of four around the vase at a spacing of 1 inch. All strands in this section are laid in parallel pairs between woven groups of four and as the size of the form has increased, the space between rows will have become slightly wider.

This last row of woven squares brings the jacket within about 2 inches of the base. The weave could have been finished in approximately the usual group-of-four weave, but the designer of the original jacket evidently wanted a less stereotyped weave, one more like that at the top. To make this weave:

Cross the inner strands of two adjacent group pairs over the outer strands in sequence and bring them together in a big elongated V figure about ¼ inch above the edge of the base. Hold with a spring clothespin.

Make four other identical V-shaped spaces outlined by inner weavers and hold them with clothespins.

The outer strands of two adjacent pairs now lie at an angle within the lower part of the V and outline a squared space. Examine to see that these strands lie correctly and weave them in sequence with the arms of the V on each side.

Make four other such groupings, weaving the strands in proper sequence.

Left and right strands from groups 1 and 3, or opposite points of a big diamond, will extend at a natural angle to meet near the edge of the base and directly below a group of four in the last regularly woven row.

Cross these correctly and hold with clothespins. If the vase is set on a table and the clothespins allowed to lie easily around it, there will be less strain on the weaver strands and less danger of breaking them.

Space all meeting of strands evenly around the base and examine the work for tautness, spacing, and errors in weave sequence.

Have ready a long strand of superfine cane or, if necessary, a strand of medium cane cut to about half of its width and well moistened.

Pull all strands down tightly and fasten in position with three rows of pairing put in tightly and the ends fastened securely. This should bring the weave even with the base.

The built-up base is made by weaving the strand ends back into themselves in much the same way bottle jackets are sometimes finished around the neck.

Working from left to right below the pairing weave, loop the left of two crossed strands, from the back, under two ends and back around the second which will also be a left end, in front of two strand ends, and back under itself with the end pointing toward the inner base.

The next loop is made in the same way by the right of a pair of crossed strands around the right end of the adjacent cross and back under its own base.

If the worker wishes to check for errors, he may remember that a left strand must pass around a second left strand before it returns to its base, and the right will go around the right end of the adjacent crossed strands. Also, each loop must surround two ends.

When all ends are correctly woven and pulled down evenly, the outer base has the appearance of a row of long-angled loops, and the base that of a flat braid.

Cut the ends of the strands to about 1 inch and fit them back neatly under the inner edge of the weave.

The two horizontal bands around the center are more decorative than necessary and are applied by running strands of cane under and over parallel pairs of strands above and below the center row of groups of four. Hold these strands in place by leaving long ends to lap back on themselves.

Oriental-weave Jacket Woven from the Top

Except that this weave (Plate 9, Fig. 4) is made tight at the top and the strands between woven groups of four are crossed right over left, it is the same as that at the top of the jacket for the vase with the hourglass form (Fig. 3). This alternate lapping of the connecting strands, rather than that of left over right will produce the same results if the sequence chosen is followed consistently.

It would be helpful to make the jelly-jar jacket before the vase jacket is attempted, for the form is smaller and will give quick practice, and the eye will be able to follow the strands in the tight weave more readily.

The jar is 10 inches in circumference and measures 2½ inches from base to neck.

A mesh of approximately 1½ inches was wanted, so by the strand count formula given earlier in this chapter, twenty spokes or ten strands of size medium cane 14 inches long were needed.

Soak the cane in cold water for fifteen minutes and wrap it in a wet towel to make it pliable.

Weave together near their centers in a small square four strands of cane in pairs, the lower right under the lower left and over the upper left.

Weave the upper right over the lower left and under the upper left. Hold with a spring clothespin.

Cross the right strand over the left strand of the pair extending on the left, and at a space of about 1 inch weave in two strands from the right to make a group of four in the same sequence as that of the first group. Hold with a clothespin.

Repeat between * for three more groups, using the ends of the first group and those of the fourth group to make the last square.

Lay this foundation around the neck of the jar and draw all strands down to tighten and center them at the cross of strands between groups spacing them evenly. The crossed strands will lie against the lower neck edge and almost parallel to it just as this part of the weave spreads out to cross the upper edge of the vase.

Cross the strand pairs from two adjacent groups and bring them together to form a group of four ¾ inch down on the crossed strands.

Put in four other groups to finish the row of small squares and form outlines of a row of triangular spaces which reach to the point of the shoulder.

The next row of groups of four should be placed just below the widest measurement and is the most widely spaced mesh. Make the small squares by bringing crossed strands together from adjacent groups and weaving them in sequence.

The final row of squares rests at the edge of the base and is 1 inch down on the crossed spokes. Hold with clothespins and examine the jacket for tautness, correctness of weave, and placing of groups.

With well-moistened superfine cane or mulberry-root fiber, put in two rows of pairing weave closely around the edge of the base, catching in the weaver strands as their positions are reached, and fasten off the pairing strands firmly. Examine the work to see that the spokes have not been allowed to loosen and that the groups of four remain in correct positions.

Finish the lower edge of the base with a small, tight coil made by wrapping the ends of a pair of right-pointing strands to the right around the rows of pairing weave, alternately, on an angle. Fasten strand ends and carry on the wrapping with other ends as necessary to complete the coil and dispose of all strand ends, trimming out ends as required for a symmetrical coil.

With the first row of angled wrappings in place, use the left-pointing strands to weave alternately, as individuals in a pair, at an angle that will make a pattern of symmetrical crosses over them around the coil.

Make the weave tight, keep the angles true, and trim out the ends as is necessary to maintain an evenly shaped coil.

Jacketed Calabash for a Lawn-party Punch Bowl

As it was designed for backyard entertaining, the rustic characteristics of this punch bowl show to best advantage in outdoor surroundings.

A very large gourd is known as a calabash and varies only in size from the growth habits of the less spectacular members of its family, such as the ladle shown, known as a long-handled gourd. The ladle was made from a gourd purposely selected for size though this variety usually grows to twice this size or more with very long handles. There are also small round gourds which make very satisfactory handleless punch cups.

The gourd for this punch bowl comes within ½ inch of being 4 feet in circumference and is the product of a northwest Florida farm. In the past, gourds were not uncommon in the rural South, where a casual hole or two was dug near a back fence, a few seed dropped, and the results left to Nature. Gourds were used in the kitchen as dippers and as containers for many products. According to popular belief, eggs kept in a gourd would not freeze regardless of the severity of the weather, if the gourd were so cut that the top could be fitted back on.

Gourds should be cured on the vine and must be allowed to dry thoroughly before being cut if the shape is to be held. This takes months for any gourd, particularly one of this size. When cured, it will be very light in weight and likely the seed can be made to rattle. Instructions as to cleaning, cutting, and waxing will be found in Chapter 12.

The oriental weave is best made with chair cane, which comes in various widths. Wide binding cane was used for the development of the jacket illustrated in Plate 1.

Using the formula for strand count given earlier in this chapter, twelve groups of four or forty-eight spokes are required for the very open mesh wanted. As the strands are woven together at their centers on the base and pass to the finish line at the upper edge, this means twenty-four strands of cane with a length of at least 58 inches to allow for the spiral, the weave take-up, the coil finish, and ease on the ends for manipulating.

The weave itself is a very simple over one, under one or under one, over one which progresses in a series of crossed spokes woven into groups of four around the form in rows at regular intervals. The cross of the spokes between woven squares is a part of the count and preserves the sequence of the weave so that the count may begin anywhere on the jacket and the rhythm will be maintained if there are no errors.

The mesh may be large or small as the dimensions of the form demand, controlled only by the number of spokes used.

For ease in handling, coil the strands individually and soak them in cold water for about thirty minutes. Wrap in a wet towel until pliable.

Following Drawing B, which shows the base weave of the glass-top jar in Plate 9, Fig. 2, with a six-sided opening, weave a base having twelve sides around an opening about 4 inches in diameter. Tighten and space the woven groups of four evenly, holding the weave in place with clothespins. This is exactly the same procedure as the hexagon base, merely requiring more strands to make a larger opening with double the groups of four.

Cross the inner spoke pairs from two adjacent woven squares and bring them together in a group of four, woven in sequence, to make the first group of a row of woven squares about 1 inch up on the crossed spokes.

Weave eleven other such groups, equidistant from the edge of the opening, to complete the row, moving the clothespins to the new positions.

Space the next row of squares at about 1¼ inches and continue to widen the space between rows so that by the time the jacket will fit over the side of the gourd the approximate size of the mesh decided on will have been reached. Make a tight weave; cane of this size does not slip readily and is hard to tighten. Make a frequent count of the weave, remembering that the cross of strands between squares has the same importance as that within the square and that the sequence must be maintained from base to finish line for a flawless jacket. Place the woven squares by measure for regular spacing.

When the weave is large enough to turn the side, it should be placed on the gourd and centered carefully. The placing will have to be watched continuously for it is not practical to depend on tying jacket and gourd together. Gourds are usually a little irregular so if the jacket is made on the gourd, the weave can be made to conform, place it on a heavy folded towel to protect the points.

Measure each row to see that an even distance from table to woven squares is being maintained. This will help prevent adjustments when the time for finishing off nears.

For an even position of the coil, adjust the last row of groups of four to lie in a row with their tops 1¼ inches below the lowest point of each scallop. Check the jacket thoroughly for correctness of weave, placing of woven squares, and tautness of strands. Pull up firmly on each strand at its angle.

With well-moistened superfine cane, put in two rows of pairing weave tightly and by measure at the place determined for the coil, catching in the spokes as the pairing progresses. Slip the ends of the pairing weavers back into the weave and fasten them firmly. Directions for the pairing weave are given in Chapter 12.

Go over the spoke strands again to see that they were not loosened or forced out of position.

To make the coil, take two of the jacket spoke ends pointing right and weave them alternately over the pairing rows in an over-and-under movement to the next group of four on the right. Slip the ends, one after the other, under the pairing rows and down into the weave behind the jacket weavers angled to the left. Arrange neatly and tightly but do not trim off excess length as all trimming should be deferred until the jacket is definitely finished and has dried.

When the spoke ends pointing right are woven, those pointing to the left should be woven similarly except that in weaving around the pairing, they should be slipped under the spoke ends already in place and at such an angle as to make a series of symmetrical crosses around the coil. This treatment gives a decorative note to the coil and slipping the ends under those already anchored will help tighten the weave around the coil. Slip the spoke ends into the jacket weave at the proper angle and pull tight.

Dry thoroughly and trim, if necessary.

Braided Ring for Base of Gourd Server

This is an adaptation of a fun-maker called a "beau-catcher" in early rural Florida but which likely originated in the Orient. It was a 6-inch or shorter section of cylindrically woven palm frond which the girls carried with them to parties, sometimes made on the spot; they would attempt to slip the front or finished end on the finger of a young man of their choice. Once in place, it holds firmly as long as the tension is maintained, though it releases readily when it is pushed from the front. Doubtless the young woman had excellent cooperation if her swain were a willing victim.

The top of the gourd shown was not symmetrical and would not hold an upright position. This fault was corrected by cementing a ring of wide binding cane, made as described, on the top to widen its base and to make it level.

If the material to be used is binding cane, as used for the punch-bowl jacket, the outer or finished surface must be kept uppermost.

The width of the cane affects the diameter of the braid, so wide binding cane was selected to make it. A wider but a less smooth and a less easily manipulated material would be flat reed.

Cut two strands of cane at least ten times the circumference of the gourd top which needs widening. Halve each strand to make two pairs.

Moisten as usual for thorough pliability.

Bend a strand at the center to make a V with the left arm on the underside but twist it so as to have the finished side uppermost.

Lay another strand in this V with the rough or unfinished side of the cane up, centers of the strands even. This exact layout is necessary.

Weave over one from the right and lay the strand parallel with the left arm of the V.

Weave under one, over one from the left to make a double V and hold with a spring clothespin. This forms the layout for half of the cylinder.

Make a second double V, tighten, and hold with a clothespin. These are to be moved along as the weave progresses until no longer needed.

Weave the left strand of one group of four, or double V, into the right side of the second double V, under one, over one and the upper left strand of the right double V over one, under one. This results in two groups of parallel strands of four weavers each, pointing left and right. Tie a string around the middle of parallel groups temporarily.

Turn the work wrong side up and bring the right outer strand up and over so it can be woven into the strands on the left under one, over one, under one, over one, and again from the right, over one, under one, over one, under one.

This is the entire procedure. Keep the weave tight by pulling on all ends frequently. Turn the work so as to present the weaver next in sequence and weave from right to left. Hold the work in the left hand with the left forefinger inserted into the upper or working end of the cylinder. The weave must be tightened continuously so as to make a very firm weave.

When the cylinder is about 1 inch longer than the circumference of the ring needed, this being an allowance for lay and tightening, hold the weave with a clothespin and remoisten to make it more pliable.

Reduce the bulkiness of the V points by pressing them firmly with pliers and slip them into the opposite end of the cylinder for about ½ inch.

Finish off the ring by weaving the strands back on themselves. Careful choice of weaver is necessary as it must be done in the exact right order if the joining is to be without flaw. Experiement until there is no doubt that the right order of weaver is established and slip all ends back over themselves into the weave, which is slightly loose at this point, lapping for three or four checks. Tighten the strands in rotation and measure for size. Hold the weave firm by bending a strand back on each side and slipping it into the weave. After the ring has dried, the other ends may be trimmed off close to the sides of checks.

For the sake of the young folks who may be interested, the finish of the beau-catcher is given, though it is the beginning which is vital!

If the medium used is palmetto or any other palm frond, which may be used green or uncured, it will be double-faced, that is, with no definite right side. Therefore, the part of the directions for turning the left arm of the weaver so as to present the outer side and laying in the second strand wrong side up should be disregarded in making the double Vs. With this exception, which simplifies the procedure, the directions may be followed to make a smooth and efficient front end. All that is necessary for the finish of the cylinder is to slip the strand ends back into the weave in any order that will hold the weave in place, trimming off the surplus frond.

CHAPTER NINE

SOME DOORYARD MATERIALS

THIS CHAPTER calls attention to some of the many basketry materials right at hand, to be had for the harvesting, giving suggestions and procedures for their use. Securing them by one's own efforts, in the home garden or elsewhere, does much to enhance their value in one's estimation.

Examples and directions for the use of honeysuckle vine, grass, umbrella-plant stems, and bamboo splints are given here. In other chapters on particular products or specific procedures, examples using cattail rush, palmetto frond, longleaf pine needles, mulberry-root fiber, and wire grass are given. Notes for harvesting and curing are given in Chapter 12.

All the materials discussed here are indigenous to a temperate climate. Most of the projects detailed here are adaptable to other materials, procedures, and forms. These instructions, while specific, are meant also to be suggestive of new creations for the imaginative worker.

Honeysuckle-vine Jacket Made with Solomon's Knot

The Solomon's-knot procedure is a very old one to which honeysuckle vine is especially adapted. It will form the necessary loose knot which will hold its position, yet allow the passage of strands through it. Actually, it is only the square knot, familiar to every Boy Scout, made over right and left strands from adjacent knots to form an allover pattern.

When the form to be jacketed is round, any multiple of four may be used for the number of strands, the size of the motif being determined by dividing the circumference of the bottle by four, with the placing falling naturally. In a rectangular shape where the motif must be centrally placed, careful planning is necessary to provide the right number and the placing of spoke strands.

The bottle in Plate 10, Fig. 3, is 10½ inches high with side panels 3¾ inches at the base and 5½ inches at the shoulder. On a bottle of such proportions, sixteen strands or thirty-two spokes were allowed since eight strands per side seemed to be the smallest number practical. These work out to a center motif and a half motif on each corner for the four panels. In alternate rows, there are two motifs per panel.

Strand length should be six to seven times the length of the form to be jacketed if the strands are to pass entirely around the bottle; however, strands only about half this length may be used if they are woven together evenly on the base.

Splicing should be avoided. If a strand breaks, begin at the point of break and trace backward into the base weave with a new half strand, removing the broken strand as it is replaced, and giving due attention to proper tension. Fasten the end securely in the base weave.

Prepare the material by coiling lightly and soaking in cold water for thirty minutes the necessary number of strands with weaver strands for the base weave, wrap in a wet towel, and allow them to lie until they are pliable.

Make a mat which will cover the base of the bottle by weaving eight strands into eight strands by pairs at their centers.

Fasten these together in an over-one, under-one weave by pairs with a weaver strand, the heavy end of which has been allowed to protrude for about 4 inches to provide the uneven number of strands necessary for a continuous weave.

Begin at once to form the weave into a square so it will conform to the shape of the bottle base.

When the weave is about 1 inch in diameter, begin to lay in the other eight strands, two at a time, in the largest spaces. When all are in, the thirty-two-spoke layout will be complete.

Separate the pairs and work to an even spacing of strands by the time the mat will cover the base of the bottle.

When the edge of the base has been reached, turn the spoke strands up closely around the bottle, centers even, and continue the weave to a depth of about ¾ inch.

Finish off the weaver and the extra spoke end by running them back into the base weave invisibly. Redampen the jacket as necessary.

Set the bottle on the jacket base, centers even, with a group of four at each center panel and on each corner.

Tie the jacket base firmly in place with small, strong cord from neck through base and back to neck on four sides.

It is recommended that the unfinished jacket illustrated in Plate 10, Fig. 3, be examined and the following instructions be mastered before the jacket is begun.

Make a row of knots around the bottle beginning with a center panel group of four strands.

Place the outer left strand under the two center strands to point right.

Pass the outer right strand over the two center strands to point left.

Bring the left strand, now pointing right, over the base of the right strand and under the two center strands to point left.

Bring the right strand, now pointing left, under the base and end of the left strand, over the two center strands, and through the loop on their right made by the left strand.

Draw down to make a symmetrical loop knot.

The center strands around which the knot was made form the right and the left side of knots in the second row, and the third row of knots is formed with the same strands with which the first row was made and placed directly above the first row knots.

This results in an allover pattern of encircled diamonds with knots at each corner. Once established, the motif regulates its own size in large measure, though attention must be given to the placing of knot rows.

Make the row of knots at the point of the shoulder the last and draw the strands straight up to the neck under a tightly tied strand of fine, immature vine wrapped twice around the neck and tied in a square knot. Spread the ends to be caught in the finishing roll. When well dampened, immature honeysuckle vine is both pliable and tough but it flattens in use which makes it ideal for this stay.

Dispose of the ends in a very small, neat roll around the neck, trimming out the ends as the strands become securely fastened in order to hold down the size. One of the longer strand ends may be used to put in a row of evenly spaced whipstitches around this roll.

Lawn Grass Used in an Open Weave

This jacket (Plate 10, Fig. 5) is an experiment to see just how Bahia grass found uncut in the shrubbery on the lot boundary can be worked up.

The material dries quickly, is not brittle, moistens readily, and, until it has been wet several times, has a waxed feel that is surprising and makes it very pleasant to work with. It is still a soft dark green though it was harvested nearly 6 years ago but of course exposure to light and time will cause it to fade.

A bottle 5 by 3 inches was chosen to illustrate the use of this grass. The jacket pattern was adapted from the weave of an imported beach slipper, made with some kind of small soft rush which required only one strand for a spoke. The length of the Bahia grass is good, as much as 30 inches, but it is fine and it tapers quickly from a base width of about ¼ inch, requiring multiple strands to make up the spoke and weaver size. Most of the width is near the root and is brown so it needs to be trimmed off. Actually, a simple over-one, under-one weave would be much more suitable for it, using either straight or shaped spokes depending on the shape of the bottle.

The base is made up of a simple over-one, under-one weave with a nucleus of four group strands woven into each other at their centers and fastened with a few rows of weaver strands, the heavy ends of which were allowed to protrude for about 4 inches to provide the extra spoke necessary for a continuous weave.

As soon as the spoke strands are stabilized, begin to add other spoke groups, two at a time, to make up the necessary thirty-four spokes by folding them at the centers and catching them into the weave. The inserting of two spokes at once does not disturb the sequence of the weave.

When the mat will cover the bottle base, center it on the bottle and tie them together with soft, strong string, neck over mat and back to neck on four sides.

Continue the weave up the outer base of the bottle for ½ inch, drawing it in snugly and equalizing the spaces between the spokes.

When the weave is even, run the ends of the extra spoke back into the weave invisibly to fasten it.

Put in a row of pairing closely around the edge of this weave from left to right. Instructions for the pairing weave are given in Chapter 12.

Turn the bottle and put another row of pairing immediately above the first. Turning the bottle will reverse the weave and give the two rows close together the appearance of a chain stitch, also adding firmness to the weave. Fasten the ends neatly and securely.

Other bottle jackets may seem to be very similar, but this weave has a distinct individuality in that the crossed spokes are the same pairs from base to neck. In other examples, the spokes ascend on right and left angles with different spokes crossing each other.

With a slightly coiled weaver strand about the size of a pencil lead, put in a row of pairing ¼ inch above the base weave, catching in the spokes, right crossed over left. Make the weave tight but with enough ease to fit the rounding form of the bottle and set a vertical position with the spoke crosses. The peg holder will be useful with this part of the weave.

Reverse the pairing to put in another row close above the first and finish it off neatly.

Put in three other double rows of pairing around the bottle with a spacing of ¼ inch

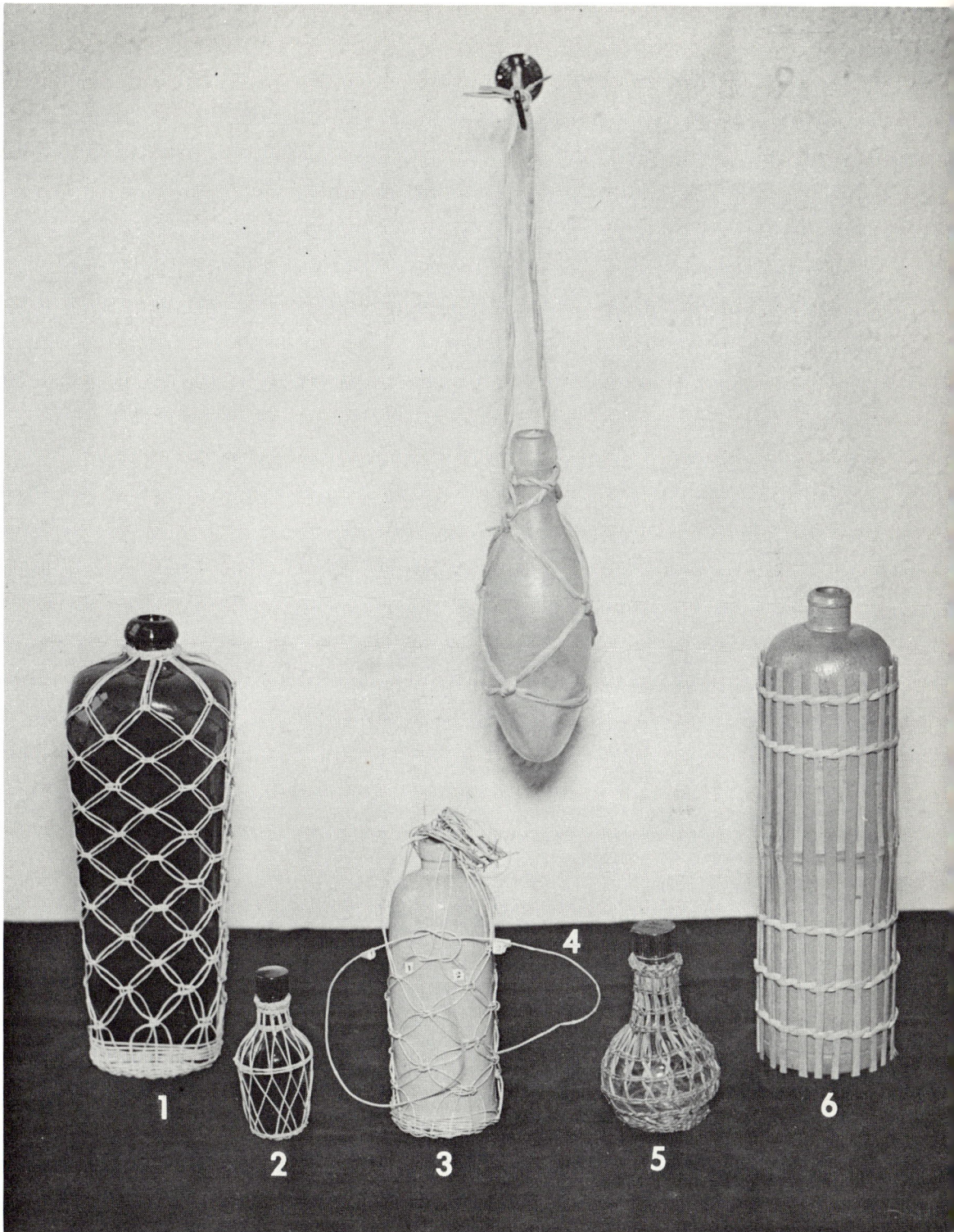

PLATE 10. Some Dooryard Jacketing Material:
Honeysuckle vine jacket in Solomon's knot (fig. 1); miniature bottle jacket of honeysuckle vine (fig. 2); unfinished jacket showing Solomon's-knot procedure (fig. 3); cyperus stem sling for pointed-tip bottle (fig. 4); open-weave jacket made of lawngrass-gone-wild (fig. 5); jacket for a pottery bottle of cane binder over bamboo spokes.

between rows, weaving in the crossed spokes evenly, right over left.

This should cover the bottle up to its distinct taper toward the neck. Put in a row of pairing, catching the spoke pairs in together but not crossing them. They will form an inverted V pattern below the base of the neck.

Reverse the weave, make the row of pairing double, and fasten it off neatly.

Weave three other double rows of pairing, equally spaced, around the neck, catching the spoke pairs into vertical single spokes. Place the last pairing row just below the bottle cap base and fasten the ends of the weavers firmly, for considerable strain will be placed on these rows when the spoke ends are run back through them.

Trim out the spoke ends as necessary to regulate the size of the strands and moisten thoroughly.

Coil a strand, bring it across the spoke on its right, behind the third spoke, and out to the front.

Repeat this procedure with all strand ends, then pull them down evenly to make a small close edge around the neck of the bottle.

Thread a coiled strand into a thin, large-eyed needle.

Slip the strand under the two protruding strand ends on its left and, by using the needle eye first, down behind itself, through the pairing weave rows, and out below the second row of double pairing from the top.

Trim off the ends closely after they have dried.

Bamboo Jacket for a Straight Bottle

The well-seasoned tines of an old bamboo lawn rake resulted in the jacket shown in Plate 10, Fig. 6.

The bottle is 11½ inches tall by 3 inches in diameter. Five rows of binding cane made the weave 1 yard for each row, finished independently.

The labor is in the preparation of the bamboo, which the worker will find to be absolutely uncompromising. Cut off the curved ends and unsplit tops, leaving the strips quite long on both ends for the sake of any necessary adjustments. The strips will be unnecessarily thick and should be split to about 1/16 inch. The width will be about right, though any inequalities should be trimmed off. The joint scar is rather decorative and breaks the starkly plain center section. If used, it must be on a line around the jacket. In the jacket illustrated, the scar was placed just slightly above the bottle's center and was accented by leaving the space free of weave.

Binding cane is heavy, and should be soaked in water thirty minutes.

After cleaning the bamboo strips thoroughly, sand the backs to the thickness desired, trimming down any bulges, determine the location for the scar, and trim any excess from the ends of a strip.

The jacket does not extend under the base of the bottle, so measurement should be from the outside base to the shoulder. Sand the lower end in a straight line and round the corners of the upper end. Using this first strip as a guide, prepare the other strip spokes identically.

The bottle used here is a warm tan which contributes to its attractiveness and makes complete covering undesirable. Therefore, only twenty-two spokes ¼ inch wide were used for the jacket illustrated. The size of the cane weaver adjusted the width spaces to a little more than ¼ inch.

Place a heavy rubber band around the bottle below the shoulder and another above the base. Slip the spokes under the rubber bands and distribute them around the bottle.

The binding rows are actually a pairing weave in which the strands are normally woven by a turn of the hand, but with the spokes held against the bottle by the rubber band, the usual method of weave is not possible. The same effect is secured by manipulating the weaver strands separately and depending on the left thumb to hold the strand tension.

Midway its length, loop a weaver strand around a spoke about ¾ of an inch above its base.

Slip the upper weaver strand under the first spoke on the right and bring it up between the first and second spokes.

Pass the lower weaver strand over the first spoke on the right and the upper edge of the first weaver strand, under the second spoke, and bring it out between the second and third spokes.

This routine crosses the lower over the upper weaver between spokes and makes a firm support around the spokes — an actual pairing. Repeat this movement with the weavers alternately around the bottle and draw tightly together.

To fasten the weavers and dispose of strand ends, bring the lower strand over the space between the loop and first spokes, pass it through the loop from the lower side, and pull to tighten. Wrap once around the lower weaver strand under the spoke first woven to leave the end lying flat and pointing up.

Pass the end of the other weaver down between the spokes in pattern to be wrapped around the pairing under the second spoke and lie pointing up. If the ends are long enough they may be caught in the next row of pairing.

When finishing off the top row of weave, slip the weaver into the loop from the top. This will put the end in position to wrap the weaver for a downward turn.

Some other suggested spoke materials are flat or round reed, any strips of constant size, native bamboo, scrub palmetto stem split to size after curing — either the flat or rounded edge may be used. A waxing of hard-surfaced mediums after they have dried will improve the appearance.

Sling for a Hanging Bottle

Even though this title puts the emphasis on the sling, much of the interest in the example shown in Plate 10, Fig. 4, is in the shape and color of the bottle, an early-twentieth-century soft-drink bottle with pointed end and thick glass, colored a frosted blue green by time.

The sling is made of four umbrella-plant stems and is visible proof of an earlier statement that craft materials may literally be found at one's doorstep. The umbrella plant was a prized specimen in old Southern gardens. Frequently today it is found in gardens and parks, and, with its near kinsman, cyperus papyrus, flourishes in low places and at the edges of lakes. For harvesting and curing instructions see Chapter 12.

Four stems of the smallest diameter and nearly 4 feet long were selected and the heads trimmed off. They do not take moisture readily, but soaking in a cold bath for about an hour, followed by a twenty-four-hour period in a heavy, wet cloth will make them fairly pliable. However, any slight lack in this respect is certainly made up by an extraordinary toughness.

Tie the four strands together in pairs in a square knot placed slightly off center. Twist the pairs together, separately, for a very few turns and bring them together in another square knot at a space of 3 inches on the long ends. This forms the first loop for the sling through which to thrust the lower end of the bottle. The space between the knots must be the same on both sides for fit, symmetry, and an even base from which to measure.

Bring the two inner pairs of slightly twisted strands together and tie with a square knot 3 inches from the first knots, centered above and between them.

Repeat on the opposite side of the bottle. These ties will outline a triangular space on each side, the base lines of which are formed by the strands which made the first loop and lie at a true horizontal across the lower end of the bottle.

Repeat on the opposite side. These ties will outline a triangular space on each side of the bottle.

Bring the lightly coiled right and left strand pairs from these last knots together and tie them near the base of the neck in a square knot directly above one of the first two knots. This will be the hanger knot and will form the upper corner of the outline for a large diamond-shaped space.

Repeat on the opposite side, placing the knot exactly in line with the first hanger knot and a measured half of the neck to secure a true hanging bottle.

Set aside the two longest strands from each hanger knot.

From one of these knots bring the other two lightly twisted strands up firmly at an angle, around the neck of the bottle, and back tightly to be fastened invisibly in its own knot.

Repeat with the two other short strands on the opposite side, crossing strand pairs between knots to form an elongated shield-shaped space below the top of the bottle.

Twist the hanger pairs with a long, easy coiling motion, bring them together above the exact center of the bottle, and tie the ends together at the desired length in a square knot.

Examine for appearance of coil and to see that the bottle hangs true.

Cut off hanger ends about 1 inch behind the knot. Trim other ends after the sling has dried.

CHAPTER TEN

JACKETS MADE WITH COILED MATERIALS

THESE FEW EXAMPLES of coil procedures, used with multiple strands or with soft fibers, will give an idea of the possibilities of the use of coils in jacketing.

Coiling will lengthen short fibers and strengthen weak ones. It will enlarge such mediums as wiregrass, pine needles, and fine grass by combining the individual rods or blades into a piece of significant size, at the same time providing an amazingly easy splicing.

Coils may be tightly twisted so they will furnish their own support or spoke, or they may be gathered into a loose, untwisted bunch of parallel strands to be fastened together and to the adjacent coils with a binder of other materials.

The loose, untwisted coil is usually made up of materials with limited use so the binder will need to be of other and more pliable materials. It may be used simply as a whipstitch at regular intervals over the new coil and through the upper part of the coil already in place, or the binding strand may be passed completely around both old and new coil with a knot between them or of more elaborate stitching which will fasten coil to coil to build up the jacket.

Coil and spoke may be of the same material, or a single spoke of other and less pliable material may be used.

Coils, in fact, are versatile and it is hoped that experiments with other materials and other combinations than those shown will be inspired by the suggestions in this chapter.

Jacket of Mulberry-root Fiber over Reed Spokes

This procedure is a variation of that used for the shuck jacket shown in Plate 11, Fig. 4. It is suggested that the instructions for making both jackets be read before beginning work on either in order to get a clear idea of the differences in procedure. In this jacket the spokes are of reed and are comparatively stiff, yet they show the tendency to drift in response to the pull of the weaver. Also, the first movement of the weaver passes it under the spoke in wrapping, which places it close against the form in its passage from spoke to spoke and makes a less bulky jacket. This weave is

planned to move from left to right showing another facet of its adaptability.

The jug measures 3 ½ inches at the widest and is 6 inches tall (Fig. 2). These measurements will require a length of 20 inches for the eight spokes of No. 4 reed to pass entirely around the jug and for ease in finishing off.

Mulberry root is a comparatively little known fiber but it is very satisfying to work with. It moistens readily, holds the moisture well, is not brittle but extremely tough, and splices with ease. The reed will require a longer moistening period in a dampened towel after a short cold-water soaking.

Locate the center of the spoke strands and bind them together with a strand of mulberry root.

Spread these spokes evenly and make a firm pairing weave at the center with a small strand of mulberry root.

Increase the size of the pairing strands to a coil slightly larger than the spoke diameter as the spread of the spokes will permit and make a mat to cover the base of the jug. When the mat will reach the edge all around, slip the weavers back into the weave invisibly by using a bag needle.

Insert mulberry-root strands for weavers which will coil a diameter of ⅛ inch on the right side of each spoke by using the bag needle and fasten the ends firmly on the inner (top) side of the mat. If necessary, these coils may be made up of smaller strands.

Tie mat and jug firmly together on four sides, from neck through mat and back to neck.

Carry the coiled weaver under, over, and back under the spoke at the right of the one by which it is based.

Repeat between * close around the outer base of the jug while making an effort to keep the central or over-the-spoke part of the wrapping as nearly horizontal as possible.

At a sharp angle, repeat this under, over, and under weave around the jug ¾ inch up on the spokes.

The size and shape of the form and the number of spokes will control the width between spokes and the length of the coiled weaver from spoke to spoke. Weavers will, to some extent, spread to conform to the shape of the form, but each row of

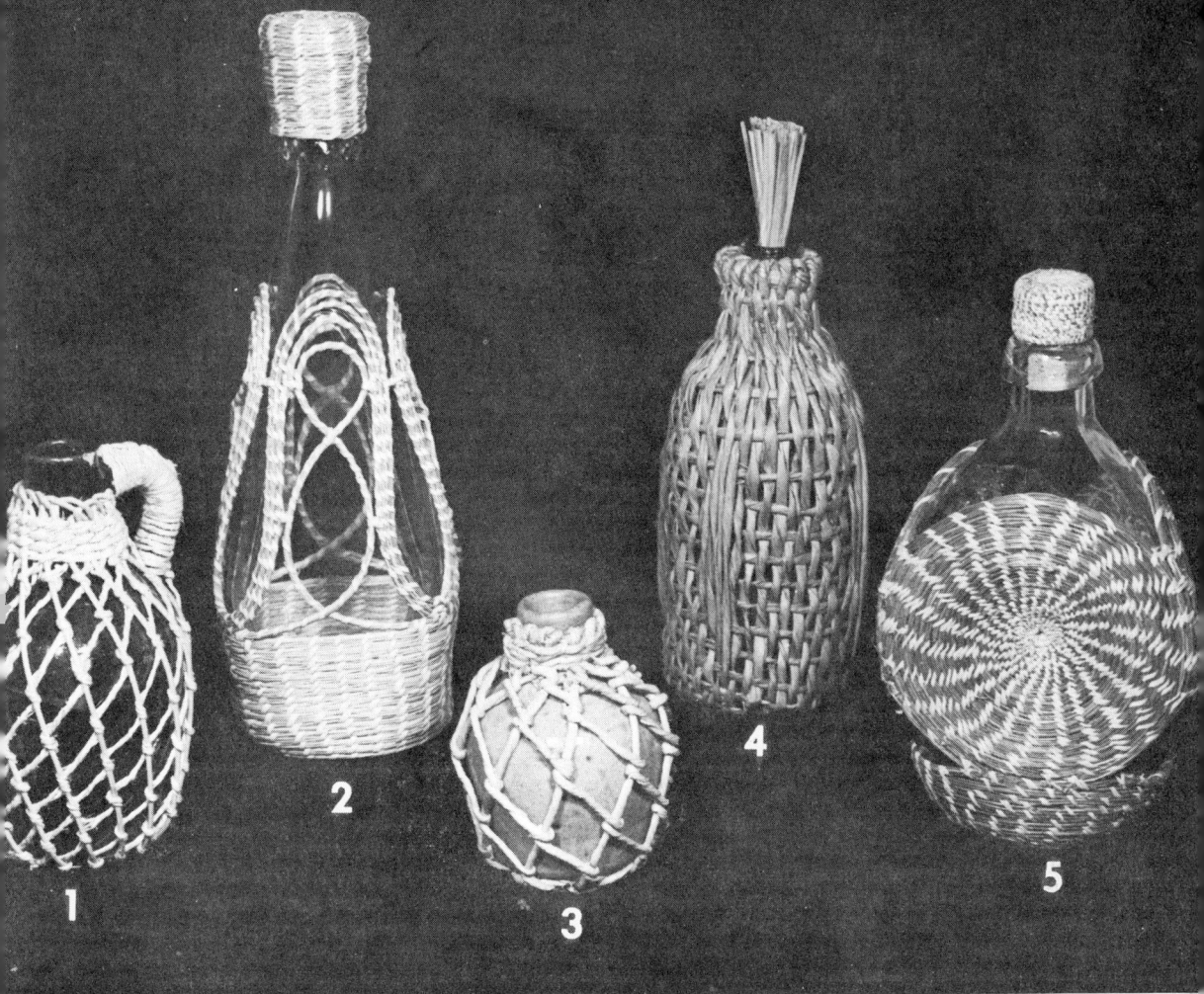

PLATE 11. Jackets with Coiled Materials:
Mulberry-root binder over reed spokes (fig. 1); wire-grass jacket bound with Palmetto frond, with whip-stitch cap (fig. 2); corn-shuck binder and spokes for a small pot (fig. 3); loosely-woven silver palm jacket with paint-brush stopper (fig. 4); pine-needle and raffia jacket for a pinch-bottle, with raffia whip-stitch cap (fig. 5).

weave should be at an even distance from the base and each weaver coil carried forward at the same ascending angle.

The spokes will require frequent adjusting and the one selected for the handle stay must be watched constantly to prevent too much drift.

Finish the jacket up to about the middle of the handle, making rows of weave to fit the contour of the jug but leaving the handle stay unwoven after the row of weave just below the handle is put in.

Tie the spokes firmly at the neck temporarily and check the work for placing and correctness of weave, spacing of spokes, and angle of weavers.

Match a weaver and the spoke on its right and with coiled mulberry root put in four rows of pairing, weaving coil and spoke as one. These rows should reach to just under the upper end of the handle and make a neat, close collar for the neck of the jug. Finish the ends of the pairing strands by running them down into the weave invisibly using the bag needle.

Fasten the end of a small, neat coil of mulberry

root at the base of the handle on the underside by twisting it in a coil tightly and pass it twice more around the handle under the spoke.

Allow the spoke to fall back into place and wrap handle and spoke three times.

Continue this wrapping of the handle, and spoke and handle, alternately until the handle has been covered to about ¾ inch from the neck.

Cut the stay spoke to 1½ inches and trim it to a long point on the underside. Turn it carefully back under itself and continue the wrapping in the usual manner except for that part of the weave made under the spoke, in which the weaver should be slipped between the spoke and its turned-back end. Finish by slipping the end of the weaver down into the pairing weave of the collar.

Trim the spokes to about 2 inches and point. Moisten well.

Make a row of small scallops around the neck from side to side of the handle by passing spoke 1 in front of spoke 2, behind spoke 3, and down into the pairing where the weave has been opened up with a knitting needle to receive it. Dispose of all spoke ends in this manner.

Push the spoke down well to form an even-edged interwoven scallop.

Jacket of Wiregrass with Palmetto Binder

The angular base, the flat panels, and the sharp taper of the bottle illustrated in Plate 11, Fig. 3, presented problems for which only wiregrass and palmetto frond seemed equal. It was also a very appropriate choice for this area as both wiregrass and the sabal or cabbage palmetto grow freely in the coastal regions of Florida, Georgia, and the Carolinas.

Normally, wiregrass requires little moistening, although it should be wiped down well for cleansing. After a long storage period, however, it becomes very dry, so it works better and is less brittle if it is given a thorough moistening.

Notes on harvesting and curing of both wiregrass and palmetto frond will be found in Chapter 12.

The palmetto frond is quite brittle when it is dry and should be soaked in cold water for about fifteen minutes and allowed to lie in a damp towel to become thoroughly pliable.

Stripping, or dividing of the fronds into strands, is best done as the work proceeds, rather than before soaking, for in the latter case the very fine strands which were prepared will seem amazingly increased in width after the absorption of water and subsequent expansion. Too, the best and longest strands are needed for this jacket as, except in the lower part of the jacket, there are few places to splice.

A coil ⅛ inch in diameter made of of twelve to fourteen rods of wiregrass, depending on the size of the rods, seemed best for this jacket. A larger coil works up faster, but it can be bulky and very hard to handle on sharp angles, for it is absolutely uncompromising.

A coil of this size will require a narrow binder strand, so the frond should be stripped to a width slightly less than ⅛ inch. The frond has a natural tendency to widen slightly toward the butt which may be controlled by taking off the widened section beginning at the butt and stripping it off by pulling toward the tip, that is, naturally, not using the stripping pin. For very narrow strands of even width, each requires this.

Thread the butt end of a binder strand into a raffia or a leather needle. The raffia needle requires an end to be turned back but if the leather needle has a tongue, the end of the strand can be kept straight and is less bulky. Also, the strand is not as likely to break at the eye as it is with the raffia needle.

Select twelve wiregrass strands or rods, group them with uneven ends and, using a long palmetto binder strand, wrap a length with ¼-inch spaces between wrappings to make a ring which will fit about ¾ inch inside the base of the form. The grouping of the ends unevenly will allow them to be incorporated in the switch or loose end of the wiregrass with less bulk.

Catch the switch to this foundation with a whipstitch over the switch, through the upper part of the foundation ring from the back with the needle placed to the left of the wrapping around the coil. Pull the strand through to the front, ready to repeat the whipstitch between the next wrapping. The strand must be kept straight, the work tight, and the angle of the stitch true to the one just below it.

The ¾-inch base weave leaves a slightly oblong open space at the center of the jacket base since the bottle is a little wider than it is thick. A base weave could be made to cover the bottle base completely, but it seemed a pity to cover the nice design at the center.

The whipstitch, made as given with the rows of stitches placed to follow the wrapping of a foundation coil or stitches in a previously laid coil, is characteristic of wiregrass work, ideally close, firm structure with slanted, even stitches placed one above the other to form rows.

Splice the binder strand by fastening the new tip in the back of the work, bringing the strand through at the base of the last stitch and over that stitch at the same angle. Pull the end of the old strand in under the switch tightly and carry it along invisibly until it is firmly fastened.

Splice the coil by inserting the butt ends of new rods into the coil as necessary to maintain its size.

When the edge of the base has been reached, ease the coil up on itself gradually, being sure that the base weave has the same number of coils on all sides of the open space in the center and that it will cover the base of the bottle so there will be no difficulty in making the jacket fit. If frequent fittings are given, the closely woven lower part of the jacket is easier to make off the bottle.

The space between stitches will have spread slightly so that by the time the edge of the base weave is reached, the spacings will be just under ½ inch. The lower part of the bottle flares so very slightly up to the angled corners between panels that the only change necessary will be to allow further slight widening of spaces between stitches.

Build the base of the jacket up to the angled corners between panels and fasten the coil with an extra stitch at the center of a corner, disposing of the weaver end on the inside.

With a long new strand fastened on the underside, begin wrapping the coil for the figure eight. Those on the front and back require a full 10 inches to the point of crossing and those on the sides 9 inches.

The first opportunity to splice comes after the

coil, for the figure has been wrapped up the right side of the lower section, around the top section, and back to the cross between upper and lower sections. Fasten the strand firmly there with a horizontal stitch which will blend with the wrappings, and a stitch on the back to hold the end of the strand.

Fasten a new strand on the back of the cross and make a vertical stitch over it. Slip the needle through the weave on the back to hold the strand in place and wrap the switch for the same measured distance for the other side of the lower part of the figure. Fasten the end at the right side of the panel to balance the beginning of the figure coil. Wrapping strands must be tight and put on evenly at a measured space between the wrappings. Those in the example shown are ¼ inch apart.

Fasten all new wrapping strands in the base weave at the centers of the angled corners. Though the size of the figures on the side panels will be slightly smaller, they should be so made as to cross, both the lower and upper sections, at equal distances from the base. This will mean a slightly more slender figure, particularly in the lower section. For uniformity, the overall length of all figures must be the same.

When the fourth figure has been completed and the coil fastened firmly at the corner intersection, rethread the needle with a new strand fastened in the base weave and wrap the switch for a coil to be turned up at an angle which will allow it to meet the lower section of the figure eight at the middle of its left side.

Fasten it to the figure with a horizontal stitch over both coils and make a tight vertical stitch between them.

Continue to wrap the switch until it will reach the middle of the upper section on its left side and fasten the coils together with a horizontal and a vertical stitch and where, likely, the strand will need splicing.

Wrap the switch for a coil length which will allow an easy arc over the top of the figure with the space of about ⅜ inch at the top between coil and figure and fasten it at the right center of the upper section in line with the knot on its left side, using the same kind of knot.

Complete the wrapping of the figure outline for a length that will balance the left side and fasten the coil firmly at the center of the panel divisions.

Repeat this outline around the three other figures.

This coil forms the base for the weave in the rest of the jacket which consists of two other coils and makes up a ⅜-inch U-shaped band around the figures. These are fastened to the wrapped outline and to each other by the whipstitch, but in order to lighten the design, the coils over the upper section of the figure from side center to side center were

wrapped only and spread slightly and uniformly.

When the last coil has been completed to the corner intersection, fasten it off on the underside securely and invisibly.

Insert the bottle in the base, fit the upper part of the jacket to the panels, and tie it firmly with soft string.

If the base has been made to fit tightly, it will be necessary to make only one set of the horizontal and vertical knots over the outer coils of the bands to secure the jacket in place. Make these where the bands come together at the middle of the upper section of the figure in line with those made to fasten the figure and the outline coils together.

The cap jacket can be made to complete the project, if desired. The cap is 1⅝ inches in diameter by 1¾ inches high and tapers sharply to the lower edge.

Make a mat which will cover the top, using not more than six rods of wiregrass with a 1/16-inch binder strand.

Make a button for the center with ten whipstitches as a foundation for the rows of stitches which will ray out to the edge of the top.

Continue the pattern down the side using the whipstitches.

The top and upper side rows may be made off of the cap but the greater part of the side weave will have to be made on the cap. A mattress needle will help with this.

Shuck Jacket for a Small Round Pot

This jar, the product of a country kiln of more than a hundred years ago, has a soft cream glaze which the color of the cornshucks seems to intensify. Doubtless its maker, who liked it well enough to scratch a script initial on its shoulder, would approve of the choice of jacketing material. It is 4½ inches tall by 3½ inches at its widest diameter. This made it a good choice for the short Florida shucks which require frequent splicing. It is shown in Plate 11, Fig. 4.

The pattern for the weave is a very old one. Dr. Otis T. Mason in his *Report on Aboriginal Basketry* calls it "wrapped weaving'" and says that markings of wrapped weaving were found pressed on ancient pottery taken from a mound in Ohio.

Moistening of shucks is the simplest of processes, for they absorb water almost instantly. Many shuck braiders keep a pan of cold water into which they dip the shuck strand as it is used. Shucks also dry quickly, so when weavers become dry, the worker should dip jacket and pot in cold water to restore the moisture.

Tie together at their butt ends nine sections of shuck about ¾ inch wide for spokes. When coiled they should be about ⅛ inch in diameter. Flatten the short end into a rosette and coil the long ends.

With another firmly fastened ⅛-inch coil, construct a firm mat in over-and-under weave to cover the base of the pot. When the edge of the base has been reached, slip the end of the weaver back into the mat invisibly by using a bag needle eye first to avoid splitting the weavers.

Adjust the spokes evenly and thread a section of shuck for a ⅛-inch coil into the mat on the left of each spoke, fastening them firmly.

Set the pot on the base mat and tie them together securely, neck through base and back to neck on four sides.

Lengthen the spokes if necessary by splicing them, tip to butt, so they will extend at least 3 inches beyond the top of the pot. Hold them in place temporarily with a rubber band around the neck of the pot.

Adjust spoke and weaver positions, coil a weaver, and wrap it once around the spoke on the left near its base at an ascending angle, bringing it from under the spoke on the upper side of itself. Tuck the end lightly under a spoke to hold it and make the same weave over all spokes at this position.

In this weave, spokes of soft material will shift in response to the pull of the weaver, but this is not unattractive and no effort was made to correct the tendency.

Put in other rows of wrapping around the pot, a row at a time, keeping weavers and spokes tight and the spaces as true as possible. Of course there will be variation in shape and size of the spaces as the size of the pot expands or diminishes.

For a pot this size, each spoke is wrapped five times by the weavers, the last time about ½ inch from the base of the neck. Then it is passed under the rubber band beside the spoke with which it will be coiled.

When all weavers have been brought to this position, dampen the jacket to have the shuck strands thoroughly pliable, wrap a strong thin strand of material around the neck twice, and tie it securely. A shuck strand would be rather bulky for this but either raffia or mulberry-root fiber would be covered by the pairing weave and would not be at all bulky.

Coil the ends of the spokes with those of the weavers brought over to them from the last wrapped spoke. The coil will likely be too large, so trim out material from both coil and weaver to make a ⅛-inch coil. After adjusting the size of all coils in this manner, put two rows of pairing weave around the neck and fasten the ends of the weavers securely. Directions for the pairing weave will be found in Chapter 12.

When all spoke weaver coils have been fastened with the second row of pairing, coil an end tightly and thread it into a bag needle.

Using the needle eye first, bring the coil across the spoke on its right, behind the third spoke, and down through the rows of pairing by the right side of the third spoke weaver. Pull the end down loosely to lie on the outside of the jacket.

Weave the end of each spoke weaver in this manner and when all are in place pull them down gradually to make a row of small, tight scallops around the neck of the pot ¼ inch below the upper edge.

When the work is dry, trim ends off close to the lower edge of the pairing.

Pine-needle Jacket for a Pinch Bottle

In the longleaf-pine area of the Southeastern states, pine needles are to be had for the picking up, which may be done through most of the year as the needles ripen and fall almost constantly. Individual pines will show differences in color, size, and length of needles, which grow in groups of three with a sheath at the end by which they are attached to the twig in clusters. The needles are cured when they fall but they will need to be thoroughly dried before they are stored. A deepening of the color may be secured by allowing them to lie in the sun for drying.

For those who do not live where the longleaf pine grows, some of the large Eastern craft-supply stores carry pine needles.

Two examples of pine-needle jackets are given, one of which is illustrated in Plate 11 and one with the miniature bottle jackets. Both use a coil of pine needles with a raffia binder but the possibilities are limited only by the imagination of the serious worker.

Various sizes of pinch bottles are to be found, some made with round panels such as the bottle shown, others with slightly oval ones. The regulation whipstitch may be used in developing jackets for both shapes, although, as the oval panel requires an oval button, the center will be more decorative if it is woven of raffia. In fact, the entire panel might be woven of raffia with an outer band of pine needles, but the panel would have to conform to the shape of the bottle panel and be made oval rather than oblong, and also slightly concave.

The pinch-bottle jacket illustrated in Plate 11, Fig. 6 is made for a bottle 7¾ inches tall, with a 2½-inch-diameter base, the 15-inch circumference being divided into three slightly concave round panels. The greater part of the jacket consists of three coil mats made to fit the contour of the bottle and fastened together at their side centers. This leaves the neck and upper section of the bottle exposed and a small triangular opening on the lower sides of the covering panels.

Moisten the raffia in cold water and wrap it in a damp towel.

Wipe down a small bundle of pine needles and wrap for dampening. These will be used to make the button and the first few rows of coil. The needles for the usual easy coils need little moistening but they should be wiped down after a careful inspection for flaws and pitch. Trim off the sheath, for that part of the needle under it is of a lighter color.

The binder may be spliced by fastening the new end in the back of the work, pulling it and the old strand to position, and fastening the old strand in place with a stitch made over it. Make the old strand secure by carrying it along tightly but invisibly in the coil.

Splice the coil as needed to keep it up to size by inserting a needle or so at a time, blunt end first.

Thread the heavy end of a strand of raffia split to a width of ¼ inch into a sharp-pointed raffia needle. This size strand will make a small, firm button with a very small opening which is desirable for so prominent a place as the center of each panel. A wider strand may be used later in the mat if preferred and the coil should be increased to twelve or fifteen needles. However, it must be remembered that once the coil is made heavier, it must be kept to size if the work is to be regular and firm.

Cut the sheaths from three groups of needles to provide nine individual needles and group them unevenly for the sake of later splicing. This will require the trimming of the blunt ends of some of the needles.

Tie the strand of raffia around the nine needles about 1 inch from the butt or sheath end and wrap tightly toward the end twelve times.

Coil the switch back on itself with the end pointing to the left and fasten end and switch together to make a very small, neat button with the end on the underside. Pull firmly on the switch to round and tighten.

With the raffia over the switch, insert the needle into the outer edge of the button between wrappings from the underside and pull the strand through to the front to form a tight whipstitch over the switch which will fasten button and coil together. Twenty-four radiating rows of stitches will be necessary in order to have them spaced at ½-inch intervals around the outer edge of the 5-inch discs which form the three side panels and, if the pattern is to be maintained, the foundation for these stitches must be laid in this row by taking two stitches between each wrap of the raffia around the coil which made the button.

In the third row, place a stitch between each stitch in the preceding coil, bringing the needle from the back through the upper half of the lower coil on the left of the stitch. Make a stitch count for accuracy.

Either a right- or a left-pointing switch may be used depending on which is easier for the worker unless there is reason for the switch to point a certain way, as in the case of the miniature jug jacket shown in Plate 12, Fig. 2, where the switch must point upward or to the right so as to be incorporated in the handle. The left-pointing switch with which the pinch-bottle jacket is made illustrates this fact.

Make the panel slightly concave by beginning at once to tighten the switch very slightly as the work proceeds. Once the trend is established, it will require little effort except to see that the fit is maintained.

When the edge of the bottle panel is reached, the jacket panel may be made to fit by tightening the coil slightly as the work proceeds and stroking it in place over the shoulder of the bottle. For a tight fit, the panels should be ⅛ inch apart at the centers of the side divisions. They will be 5 inches in diameter at this point and should be completed at the lower side about ¾ inch to the right center lower edge.

When two other identical panels have been made, bring the side centers together over the centers of the bottle divisions between panels, matching the raying rows of stitches, and sew them together firmly with stitches in line with five raying rows.

Fasten the jacket in place temporarily with strands of raffia run through the centers of the buttons, pulled tight, and tied evenly and firmly around the neck of the bottle.

Wrap a switch where it was left at the lower right center of a panel for about 2 inches to complete an equal-sided triangular space between panels and fasten the end of coil and binding strand under the edge of the mat on its right.

Fasten the second mat to the third in the same manner. These wrapped coils join the panel mats together and make a continuous coil around the base possible.

The switch at the right center of the third panel is now the only loose end and is used to carry the work forward. Wrap it to complete the third equal-sided triangular space, but instead of fastening it off as was done with the switches at the base of the other two panels, proceed with the base weave using whipstitches in line with those which made the panels. The work from this point will be easier if the bottle is turned upside down, for the jacket is finished on the bottom. The mattress needle will be helpful in making the rest of the jacket.

When the base of the bottle has been turned, build up a band of 1 inch or less, fitting the weave closely to the bottle. Finish it off by diminishing the size of the coil to leave a circular space at the center of the jacket base.

If the jacket is slightly loose at the top after it dries, it may be tightened by stroking the panels firmly upward to place them and stitching across one other pair of raying rows to draw the panel edges close together.

The jacket for the cork is made with a button and whipstitch over only six needles, although a larger coil could be used if preferred. After the top has been finished and the edge turned, the jacket will be made on the cork, so a mattress needle will be almost necessary if a tight fit is to be secured. Finish off the cork jacket at about 1 inch.

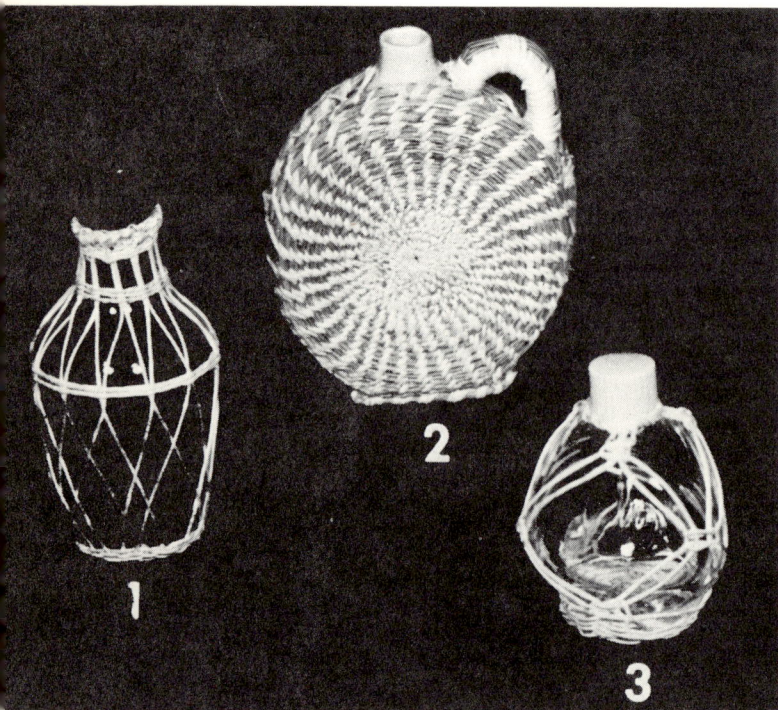

PLATE 12. Miniature Bottle Jackets: Honeysuckle vine jacket for a miniature demijohn (fig. 1); pine-needle jacket for a miniature jug (fig. 2); Solomon's-knot jacket for a miniature pinch bottle (fig.3); wire-grass jacket for a miniature water jug (fig. 4); open-weave jacket of fine

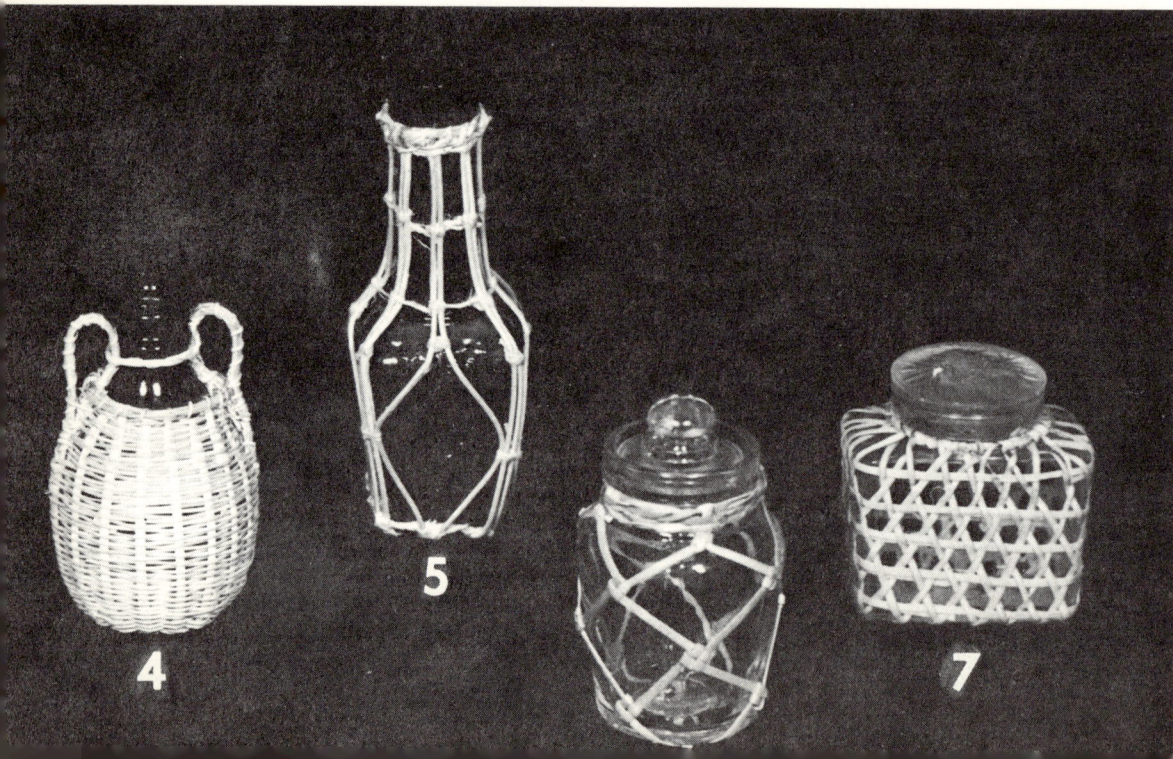

CHAPTER ELEVEN

MINIATURE BOTTLE JACKETS

THERE ARE a great many small, attractively shaped bottles on the market today that are difficult to discard when the products they contained have been used, yet as empty bottles they are rarely sufficiently decorative to retain. It was with this in mind and with the hope of inspiring experimentation by others that the examples of jackets shown in Plate 12 were developed.

Almost any procedure for jacketing larger bottles may be used for miniatures if it can be reduced or otherwise adjusted to smaller-sized patterns and materials. The work may be more tedious and may require more exact workmanship than the larger bottle jacket, but it is rewarding.

The two prime binding materials among others for the smaller patterns are mulberry root and raffia because, even when split into very narrow strands, both are strong, dampen readily, and have a satisfying length so that splicing may be held to a minimum.

Very small reed, honeysuckle vine, Virginia creeper, palmetto frond, narrow grasses, pine needles, wiregrass, and grain straw are a few of the mediums which adapt themselves to a small form.

Honeysuckle-vine Jacket for Miniature Demijohn Shape

The dark brown color and the shape of this bottle make an excellent background for the weave of very fine honeysuckle vine corresponding to size 00 or smaller reed (Plate 12, Fig. 1)

The bottle is 4 inches high with a 2-inch shoulder and a 1-inch base diameter. The spoke strands must be 14 inches long and should be graded and very fine. The length will allow the use of shorter ends left from other work for the twelve spoke strands which pass entirely around the bottle. Longer strands will be required for the weavers.

Soak both weavers and spokes in cold water for about fifteen minutes and wrap in a wet towel until thoroughly pliable.

Fasten the spokes together at their centers, six over six, by binding them with the weaver. Use the heavy end of the weaver and let the end extend on the left 3 inches in order to form the uneven number of spokes necessary for a continuous weave.

Put in two rows of weave over four, under four, weaving the extra spoke by itself.

Separate the spokes into pairs by weaving over two, under two in the third row and weave tightly and smoothly to within ⅛ inch of the edge of the bottle base.

Finish the mat to the edge of the bottle in a routine of over one, under one, spacing the spokes evenly.

Tie base mat to bottle with small strong string from neck, over base mat and back to neck on four sides, being sure centers are even.

Put in five rows of weave tightly around the lower edge of the bottle and run the weaver and the extra spoke back into the base weave and the mat to fasten them. Trim off excess length but leave close trimming to be done when the jacket has dried.

Divide the spokes into six groups of four spokes each.

*Number the spokes in a group 1 through 4 beginning at the left.

Lay 2 to the right over 3.

Lay 3 to the left over 1.

Lay 4 to the left over 2 and weave under 1.*

Secure with a rubber band below the shoulder and repeat between * for the other five groups.

This weave makes a row of diamond shapes in the groups of four around the base of the bottle. Draw spokes tight to leave a slightly wider space between groups at the base.

Weave spoke 4 over spoke 2 in the group to the left, having them meet at the shoulder.

Weave spoke 3 under spoke 2 and over spoke 1 in the group to the left, meeting at the shoulder.

When strands of all groups have been woven in this manner, the sides of the bottle will be covered with a series of diamonds, triangles, and parallelograms outlined by the spokes.

With very fine honeysuckle weaver, put in two rows of pairing just below the shoulder where the spokes cross. Check to be sure the sequence of weave has not slipped, that all right spokes cross over left ones at the shoulder, and that the spacing is even. The weaver will be so small and the space between crossed spokes so great that an extra turn of the pairing between them will be needed.

Finish off the continuous two rows of pairing by weaving the ends back over themselves neatly and firmly.

Combine spoke 4 with spoke 1 to make a series of triangles around the upper shoulder and hold in place with a tight rubber band.

This will leave all of spokes 3 and 2 which will fall naturally in place between each 4-1 triangle; however, they will not be from adjacent groups.

Put in three rows of pairing very tightly around the base of the neck and finish off neatly.

Put in three other rows of pairing around the neck just below the cap base, catching in the pairs of spokes so they lie vertically and side by side.

Finish off the jacket by making a small neat roll with the spoke ends whipped down with a honeysuckle weaver strand. Trim out any extra length as soon as the spoke is fastened in or the roll will become too large.

Pine-needle Jacket for a Miniature Jug

This jacket, made in four parts, with an applied base was made to test the adaptability of pine-needle work to a small jacket. (Plate 12, Fig. 2).

The form was chosen deliberately, although a regularly shaped bottle, possibly a bulbous one, would be easier to jacket as it would have no sharp angles and the continuous coil would require no fitting together.

The form is 3 ½ inches high by 2 ¾ inches at widest diameter and is made throughout with a coil of four needles. Of course such a small coil goes slowly but the jacket will look clumsy if a heavy coil is used.

One other example of pine-needle jackets is shown in Plate 11, Fig. 6. Reference to the general observations in those instructions and to the notes on pine needles in Chapter 12 will be helpful.

The raffia stitching in pine-needle work may be intricate and ornate or, as in this jacket shown in Fig. 2, very simple. It is actually a whipstitch made over the new coil by inserting the needle which carries the raffia through the upper part of the previous coil from the back at the left of each stitch and pulling the strand through to bind the new coil firmly in place. This method conceals the raffia between stitches and leaves only the stitch on the face of the work showing.

Pine needles require little moistening, except that given by a good wiping down to cleanse them, unless a very sharp angle is to be turned, as at centers, when it is advisable to soak a few of them in cold water for a short time and wrap them in a damp towel to lie overnight.

Select a few needles, wipe them well, cut off the sheaths, and discard any with flaws or which have resin on them.

Raffia comes in long strands so it does not require frequent splicing. It is very strong even when split fine and dampens readily, making unnecessary long soaking. A cold-water dip followed by wrapping in a wet towel for a short time is adequate. It may be remoistened as necessary.

Thread the heavy end of the raffia in a sharp-pointed raffia needle, tie the end around four pine needles about 1 inch from the blunt or sheath ends, and wrap the needles toward the short end sixteen times.

Turn the switch, or long end, back over the tied ends, pull down to make a small tight circle, and sew through both coil and switch, leaving only a very small opening at the center. The blunt ends of the needles will protrude at the left side of this button to be trimmed off later.

Enlarge the button by inserting the needle from the back through the button between each wrapping stitch and pulling the switch down firmly.

The circumference of the jug shown required thirty-two stitches at the outer edge. In order for the rows of stitches to be continuous and ray out from the center of the pattern, the foundation must be laid at the center.

To the sixteen stitches already on the button, add an extra one between each stitch, adjusting them evenly. Count the stitches for accuracy.

Splicing the switch of pine needles is done by inserting a needle, blunt end first, into the coil as needed. Splice the raffia by taking the new stitch over the last old one and carrying both ends tightly along between the coils. If preferred, the end of the old binder and the new may be fastened together on the back but the new binding stitch must be taken by inserting the needle where the last stitch came to the front if the pattern is to be maintained.

When the mat begins to extend very slightly over the side of the jug all around, decrease the size of the coil to finish off the circle smoothly.

Make a second mat, arranging for identical rows and spacing of stitches.

Wrap four needles tightly at equal intervals to those on the edge of the mat, providing thirty-two spaces, and bring it together into a circle which will fit the side of the bottle and the slightly built-up base. The pine needles are too stiff to allow the band to follow the shape of this base so it is covered up entirely and a base provided by an applied roll.

Using the ring as a coil, make a band of four coils which will cover nearly half the width of the side, placing the joining of this ring at the base of the jug. Make the rows of stitches straight across the band rather than at an angle and have the switch so placed when the band is completed as to have the end pointing upward on the last row at the base of the handle. Some of the needles will be used in making the decorative handle.

Proper sequence of stitches and the placing of the switch at the handle base require a definite right and left band and this must be planned for when the ring for the second band is made. The stitches must also be the same in number as those on the edge of the mats, and with the same spacing as nearly as possible, for the rows of stitches are meant to ray out from center front, across the sides, to center back. Reversing the procedure and working from right to left for the second band, with the needle inserted at the right, instead of left, of the stitch in the previous row of stitches should preserve the sequence and leave the switch properly placed at the base of the handle.

Fit the front mat to its band with the switch at the base of the handle and pin band and mat together with rows of stitches meeting.

Sew these together with a whipstitch so placed as to make the rows of stitching continuous.

Fasten this half jacket securely to the jug with the switch at the base of the handle and fitting tightly in place on front and side, using a small but strong, soft string.

Pin the other mat and half band together in similar manner and try for placing of the switch. Make any necessary adjustments.

Sew the back and side ring together, placing the whipstitch as necessary to preserve the pattern.

Begin at the base of the handle and pin the two side bands together at the center, matching rows of stitches. The fit should be snug and require a little drawing together which was planned for when the bands were made just a trifle narrow.

Using a mattress needle, the curve of which will allow stitching on a flat surface, fasten the sides together with whipstitches in line with those which made the bands. End the stitching on the opposite side of the jug in line with the handle base and fasten securely.

Cut and remove the string which held the half jacket to the jug.

At the end of the stitching, a long V-shaped opening was left on each side of the neck. Close this with an evenly spaced lacing from the tip of the V to the neck, around it with the lacing strand, and back to the tip with stitches on the opposite side of the opening, which will give the lacing. Fasten the end securely.

Close the opening under the handle in the same manner.

Cut out the pine needles at the base of the handle to the four best ones and make a band of three stitches tightly over the needles at the base of the handle.

Wrap the handle under the needles for ¼ inch. Wrap handle and needles three times.

Repeat between * to cover the handle, having the last wrapping covering the ends of the trimmed pine needles where the handle is attached to the jug. Fasten off the raffia strand securely and invisibly.

Wrap four needles to make a finished ring of five inches to form a base.

Shape it gently into a rectangle to fit the base of the jug and sew it in place with whipstitches in line with those on the edges of the front and back mats.

Wiregrass Jacket for a Miniature Water Jar

Wiregrass is well named as it is stiff and wiry even after moistening and does not lend itself to close turns, so other materials are used to bind the strands together. The traditional binder is palmetto frond, which grows in the same general area and complements the color and weight of the wiregrass.

In the small bottle, turned into a jar by the jacket weave and applied handles, in Plate 12, Fig. 4, the base, spokes, handle bases, and binding material are all from the sabal or cabbage palmetto.

The form is 3¼ inches high with a 1½-inch diameter base which expands to 2 inches at the widest diameter. This requires eight strands slightly less than ⅛ inch wide for spokes which should be 9 inches long as they pass entirely around the bottle. Provide weavers of the same width for the jacket base and the handle bases. Binder strands are furnished by ribs from the frond. Do not confuse the hairlike filament between fronds nor the frond midrib with this rib, which is the outer edge of each double frond. It is strong, pliable when wet, and can be stripped to about 1/32 inch, making a much better binder than stripped frond for this jacket.

Soak the frond, six or eight ribs, and a dozen or so strands of wiregrass in cold water for fifteen minutes and wrap in a wet towel to lie overnight.

Weave four spoke strands into each other to form the center base.

With a long weaver strand, the heavy end of which extends 4½ inches to provide the uneven number of spokes for a continuous weave, stabilize this button by weaving closely around it twice.

Lay the other four spoke strands across the back of the button, two at a time, and continue the weave until it will cover the base of the jar, spacing the spokes evenly in an over-one, under-one weave and making a tight, firm weave. When the base mat is a true circle, run the weaver back into the mat invisibly with a leather or raffia needle.

Lay this base under the towel to keep it moist and prepare the two handles by laying 3½ inch strands of wiregrass side by side on a 4-inch strip of palmetto ⅛ inch wide and fastening them in place with a firmly anchored palmetto-rib whipstitch. Fasten the end of the rib tightly. By having the palmetto strand longer than the wiregrass, fastening the rib is made easier and the ends of the handle are slightly less bulky.

Lay the handles under the towel to keep moist.

Tie the jacket base and the form together, centers even, with small strong twine from neck, over mat, and back to neck on four sides.

Slip the end of a single blade or rod of wiregrass into the base weave on a long slant and weave spokes and wiregrass together in an over-one, under-one weave, fitting it tightly around the form. Splice the grass weaver as necessary by laying in a new strand and carrying it along with the old weaver. Examine frequently for tightness of weave and spacing of spokes.

When the weave has passed the point of greatest diameter, or about 1 inch below the base of the neck, insert the ends of the handles into the weave over spokes opposite each other. Only the palmetto base will slip under the weave, but that will hold the handle if care is used. Continue the weave to within ½ inch of the neck base, allow the handles to stand erect, and finish the weave without including them.

Turn all spokes except the two under the handles back over themselves in this manner to hold the weave in place, as it is inclined to slip upward on a diminishing form. A ½-inch lap will be sufficient and the ends may be brought to the surface where they can be trimmed off closely when the jacket has been finished. Cut off spoke ends under handle closely.

Make a very small roll around the top of the weave by wrapping a strand of wiregrass four or five times around it and fastening it to the jacket with an evenly spaced whipstitch put in with palmetto rib.

Turn the handles back under themselves and fasten them down below the roll with a few stitches of palmetto rib made in line with the weave.

With a slightly smaller roll made in the same manner, bind the inside loop of the handles to the jar just below the cap base.

After thorough drying, trim off ends closely.

Open-weave Jacket of Very Fine Reed with Mulberry-root Binder

The four-petal conventional flower made in this bottle and placed very exactly on each of the four side panels could not be ignored, so the design literally built itself around them.

The bottle, shown in Plate 12, Fig. 5, is 5 inches tall, 1½ inches at the shoulder, with 1¼-inch base diameter. This required a 13½-inch strand length for the eight strands which pass entirely around the bottle to form the sixteen spokes of finest reed. Mulberry root stripped fine was used for binder.

Moisten both spoke and binder material by soaking fifteen minutes in cold water and wrapping in a wet towel to lie until pliable.

Cross four of the pairs at a right angle to form an X for the corner spokes and bind them at their centers.

Place a pair of strands vertically and another pair horizontally across the first four strands and fasten all together near their centers.

Put in weave around these pairs to about ½ inch diameter using only enough binder to hold the spokes together firmly. Miniature bottles have little weight and if the center of the jacket protrudes to make an uneven base, the bottle will tip.

At about ⅛ inch farther out, put in a row of knots from spoke pair to spoke pair around the center base of the jacket by tying knots over each spoke pair in the following manner:

*Thread a small-size raffia needle with the binder which has a loose single knot in its end.

Make a loop around a pair of spokes with the binder and back through the knot.

Pull the knot tight and center it under the spoke pair.

Pass the weaver around the pair of spokes from the right a second time, bringing the needle up between the spokes on the upper side of the binder loops and down between the spokes below the loops.

Repeat this movement around the binder loop a second time to make a firm close knot.

Insert the needle through the knot from the left on the underside to the right side of the next pair of spokes on the right.*

Repeat between * to make a ring of knots around the center base.

When the eighth knot has been put in, run the weaver as usual to the pair of spokes at its right and fasten the end under the knot.

The knots throughout the jacket are made and finished in this manner. In making these knots, it should be remembered that there is no standard mulberry-root strand, only one's own idea of fine, medium, or heavy, since it is not a commercial product. If the suggested twice over and through the spokes does not seem sufficiently heavy, there is no reason the wrappings should not be made the third time.

Put in another row of connected knots around the lower edge of the bottle, following the outline closely. The mulberry weaver should be made slightly heavier to compensate for the build-up at the center base.

Put a rubber band tightly around the spokes at the bottle-neck base to hold them in place temporarily.

Draw adjacent center spokes on right and left to a corner pair and make a loop knot over the four spokes parallel to the points of the flowers, keeping the spokes flat, side by side, and with the strand crossing the binder loop between the corner spokes only. Fasten the weaver firmly under the knot.

Make identical loop-knot pairs on the other three corners and fasten.

Draw the inner spoke pair back together and fasten with an independent knot at the center side panel on the point of the shoulder. Follow with knots on the other three sides.

Place independent knots on the four shoulder points.

Place a row of connected knots, evenly spaced, around the base of the neck very tightly to hold the spokes in position. Fasten the weaver securely.

At the midpoint of the neck or about 1 inch up on the spokes, make another row of connected knots, spacing equally, and fasten the weaver firmly.

Put in two rows of pairing weave about ¼ inch below the cap base and fasten the ends securely. For pairing weave, see Chapter 12.

Make a close roll around the neck just above the pairing by bending the left spoke of a pair over its own right spoke and behind the pair on its right. Cut out the right spoke even with the left spoke and take even whipping stitches to bind the left spoke end down, trimming it off after the next spoke has been bent down. If the pairing weaver end happens to be long enough, it will serve nicely for this whipping stitch; otherwise, insert a new strand.

Even with due care, the weave on the bottom may be a little thick. In this case, moisten the jacket and tap the center lightly with a hammer to flatten it. Drying the weave under a press with a coin at the center to give extra pressure at that point is also helpful.

Pigeon Box in Allover Pattern

The regulation pigeon box is made by manipulating the four ends of two strands. The jacket on the glass desk jar illustrated in Plate 12, Fig. 6, shows the possibilities of the procedure when there are only two loose ends to a box to carry on the weave. Other examples are given in Chapter 6.

The jar is 2½ inches high with a 1¼-inch base and a 2-inch shoulder diameter. This will require four strands of size fine-fine cane 22 inches long. Coil the cane loosely over the hand and fasten it. Soak it in cold water for fifteen minutes and wrap it in a wet towel until pliable. Dry cane is brittle and will break when it is creased sharply so it must be kept damp while it is being used. Redampen it as necessary.

Make a regulation pigeon box for the center base of the jacket by bending a strand at the center to form a V with the left end on the underside.

Lay another strand in the crotch of the V with its underside up.

Number the strand ends from the left, 1 through 4.

Lay strand 4 to the left over strands 3 and 2.

Lay strand 2 over strands 4 and 1.

Lay strand 1 over strand 2 and through the crotch of strands 3 and 2.

Pull all ends to tighten and form a woven square or a pigeon box.

Slip the two other strands into the weave on the back of this square on the diagonals and center them. These two strands have no anchor and they will slip from position until the first row of boxes on the edge of the base is made, so they require careful attention until then.

The success of this jacket will depend on a tight weave and accurate measurements. To make the latter possible, provide a strip of thin, stiff cardboard and cut in at ½ inch, ¾ inch, and 1 inch, or, preferably, make a separate small measurement board for each.

With the pigeon box polished side up, proceed as follows:

*At ½ inch from the outer edge of the center box, measure a pair of strand ends and crease them lightly downward over the gauge.

For these strands use one of the box ends and the nearest end of an inserted strand. Box ends are fixed; inserted strands will slip.

Slip the loop so made in the right strand over the loop made in the left strand.

Hold loops in position and bring the end of the right strand to the right horizontally and back to the left so that it may be slipped through the left loop. Pull through, leaving a slack loop.

Bring the end of the left strand between the strands at an angle, under the lower side of the right strand, up over it, and through the loose loop in it.

Pull the right strand end to tighten the upper part of the box and follow with the left strand to tighten the lower half, square the box, and form a loop connecting box to box.*

Repeat between * to form three more boxes near the edge of the base.

At ¾ inch beyond the outer edge of the boxes, put in a row of boxes to lie ¼ inch above the base of the jar, using the upper right and left strands of adjacent boxes and the same procedure.

Measure and crease, at 1 inch strands from adjacent boxes to make the third row of boxes around the jar which should bring the jacket about to the midpoint of the jar.

Put in the fourth row of boxes by measuring strands at 1 inch again. This will bring the jacket to the point of the shoulder.

Except for an occasional trying on, the jar has not been used, but from this point on the jacket will have to be made around the jar.

Measure strands from adjacent boxes at ⅝ inch, crease lightly, and with the right strand tie a knot like a man's tie knot around the left strand by passing it over the left twice, up between the strands under the two wrappings, and tightened to lie one strand above the other at the edge of the neck. This will reduce strand ends and fasten the right end firmly but the left end will slip to make any necessary adjustments.

Pull all knots into place with the left strands and bend the four pairs down, leaving a small loop at the top. Hold with a rubber band.

With a well-moistened length of cane, put in two tight pairing-weave rows around the neck of the jar. The rubber band is not needed after the first row has been put in.

Put in the third row of pairing, but in doing so, slip the upper weaver through the loops left in the double strands and the lower one under the strands. Fasten the pairing weavers firmly and run the ends back under the weave with a curved needle.

Adjust the strands in the four loops to lie exactly one above the other and pull them down tightly. The weave around them in the last pairing row will serve to help fasten them and they will hold the pairing in place. When the jacket has dried, the ends may be trimmed off safely just below the edge of the first pairing row.

Hexagon-mesh Jacket for a Miniature Rectangular Jar

The jacket shown in Plate 12, Fig. 7 is the same weave as that on the jar illustrated in Plate 4, Fig. 3, in the chapter on palmetto weaves. Though the material for this jacket is also palmetto frond, the strands are little more than half as wide because the form is only 2 by 2 by 1¼ inches, and therefore is less able to resist tension.

Other features which make this jacket more difficult than that for the larger form are the rectangular jar with wider shoulders on the sides, different width in front and side panels which requires that the pattern be specifically placed, and the fact that the spoke strands must be carried over the shoulder of the jar to be finished at the base of the neck with the ends disposed of there instead of being folded back on themselves and down into the weave just below the cap base, as can be done with the round jar. All of these factors make this jacket a project definitely not for the inexperienced worker.

The hexagon-mesh-base instructions are given for the benefit of any who may want information on this particular point. However, a much less involved base can be fashioned in loose check weave with the number of strands required for the sides of the form. This will mean that strands lie at a right angle to base and sides, but this can be corrected by turning them over at the edge of the base to give them the proper angle.

If a round bottle is not to be used for the form, a square bottle is recommended, one with space on the shoulder around the neck more evenly divided. This will allow the combining of strand ends in pattern over a stay and a finish with or without a collar coil.

There are jacketed lotion bottles to be seen frequently in men's toilet goods departments which have this same general weave and seem very simple. An examination will show that a very long strand is used to lay on spokes at a slight angle so that the upright strands are continuous and the second wrapping is made on the other angle to form the Xs. This results in no more than four ends (possibly only two), and these are lapped back on themselves to be held with the horizontal weave of independent strands which is identical with the cross weave of these instructions. Owing to the shorter length of the palmetto frond, spoke strands of palmetto cannot be put on in this manner.

Prepare twenty strands of palmetto stripped to 1/16 inch in width and at least 10 inches in length. Plan to use the material at the base of the frond, as it is stronger. Ten inches allows for extra length to compensate for any broken strand. Also prepare one or more palmetto ribs for a stay with which to tie down the spokes at the base of the neck when the side weave has been made. This rib can be as narrow as 1/32 inch, but when it is well moistened it is very pliable, strong, and the same color as the jacket. This is the rib on the outer edge of the double frond, not the frond midrib. See Chapter 12 for directions for harvesting, curing, and stripping palmetto frond.

Soak the prepared material for fifteen minutes in cold water and wrap it in a wet towel until pliable.

This weave, which must always be made with an even number of spoke strands, is an allover pattern of hexagon spaces in rows outlined by Xs made by crossing strands at one angle consistently over those on the opposite diagonal. The example given in Plate 4 shows the right-to-left angle crossing over the left-to-right ones, but in this jacket, the left-to-right strands cross over the right-to-left ones, proving that the lapping need only be consistent. The crosses are fixed in place with independent weavers in the crotches of each row horizontally, over the upper right arm of the cross and under the upper left arm or over the over and under the under spokes. This weaver is subject to no strain and will hold its position if the ends are lapped about 1 inch. Do not trim closely until the jacket has been finished and is dry.

Place a heavy rubber band around the jar near the base and slip the ends of the four horizontal strands under it at both ends across the base, centering the strands and spacing them.

Run the ends of several of the front-to-back

strands under these horizontals on a lower right to upper left side at an easy angle across the base and under the rubber band. Center and space these strands, pulling them down tightly. The spacing should be about equal to the width between the horizontal strands. Establish the angle across the base with these first few strands and fasten them in position with strands on the other diagonal woven over the horizontal and under the diagonal across the base. If the first rows of weave are properly put in, this second diagonal can be woven from either the right or the left side of the base with the same results.

Fill in the space on the base with the rest of the spoke strands arranging the X pattern to outline a true hexagon space and so placed as to have six Xs in front and back between corners. Examine for correct sequence and spacing.

The weave up to this point has been made entirely by the lower left to upper right strands and is used for the base only. The first horizontal and diagonal rows are merely laid on and held by the rubber band so the third layer of weave is used to stabilize the first two. On the sides, this duty will be performed by the independent horizontal strands.

Tie the woven base firmly to the jar with strings from neck, over base, and back to neck on sides and ends, centering carefully.

The arms of adjacent Xs will cross the spoke strands so that they will require rearranging after every horizontal weaver is put in, but this allows the weaver to be laid in the upper half of the Xs so that the weaving of the horizontal requires little other effort except tightening.

Lap the ends, leaving a short extension on the outer one to help with later tightening. Rearrange and tighten the spokes, pulling up on them at their individual angles. Hold the weave in place with the rubber band.

After the fourth row of horizontal weave has been put in, pull up again on all strands firmly while pushing down on the horizontal weavers to shape and size the mesh. The results will be surprising.

The fifth row of weave is the last on this jacket; however, the number of rows is determined by the size of the base mesh which should be the pattern for that on the sides.

Tighten the weave after the last horizontal weaver is put in, rearrange the spokes, and move the rubber band up over the last row of weave. It is so near the shoulder of the jar that it is inclined to slip, but the crossed spokes and the rubber band will hold it in place.

Soak jacket and jar in cold water and wrap in a wet towel for the ends to become thoroughly pliable. The finish puts a heavy strain on the strand ends.

The neck on this jar is very short so that help was needed to hold the stay in place. This was furnished by the tied-down glass top (which merely sits over the neck flange), so that the stay could not slip up. A screw cap would have been easier to handle. Remove the ties which hold the jacket in place as they are no longer needed.

Place the strand ends in order around the neck and tie the stay around them in a square knot after passing it twice around the neck. Spread the stay ends to be incorporated in the coil.

Pull up on all strand ends tightly, arranging them so they hold the cross above the last horizontal weaver.

Begin at the center front or back with a pair of strands which lie in a crossing position just below the neck and, with a long strand end, make a tight-angled stitch around the tie and a strand end to hold it in position.

Pull up firmly on the next pair of strand ends and catch an end in the coil at an angle which fits it. When ends on one angle are fastened into the coil in this manner, weave those on the opposite angle similarly. Strands should not be pulled out of their natural angle but should be made to lie smoothly in the coil.

Do not attempt to use all of a strand end but trim it out when it has been secured, in order to hold down the size of the coil. When a strand end which is being used for a binder becomes too short, fasten it down with another which will take its place. Should all strands be too short to be used for binding, provide a well-moistened extra strand for this purpose.

The sides have wider spaces on the shoulder but they can be placed and fastened so as to give a light lattice effect on the side shoulder. Any lapping back makes this part of the weave seem cluttered so these side ends must be disposed of in the coil also.

Trim off any ends after the jacket has dried.

CHAPTER TWELVE

GENERAL INSTRUCTIONS

IN THIS CHAPTER will be found instructions for procedures used frequently, some harvesting and curing directions for those who would like to provide their own materials, suggestions for the washing of jacketed forms, and the instructions for cleaning and waxing gourds.

Given are the directions for making the pairing weave, slip-top knot, and four-strand round braid.

Harvesting and curing directions are given for bamboo, cattail rush, cornshucks, honeysuckle vine, mulberry-root fiber, palmetto frond, pine needles, umbrella-plant stems, Virginia creeper vine, wiregrass, and yard and other grasses.

Directions are also given for washing jacketed forms and cleaning and waxing gourds.

The Pairing Weave. The pairing weave is a frequently used basketry procedure and is employed in developing various projects given in these instructions. The name describes it accurately, for it is a process in which two lightly and independently coiled weavers are used to hold the spokes in place. This twisting or coiling of the weavers goes on constantly between the crosses they make around the spokes. For symmetry as well as to make adjustments possible in spoke placement, both the coiling as well as the crossing of weavers must be done in the same direction. Preferably, this is forward, down, back toward the body, and up, making a continuous gesture which leaves the weavers in place ready for the next coiling. A little practice with a pair of weavers around the fingers of the left hand will make the movement clear.

Pass a lightly coiled weaver around a spoke, off center for the sake of future splicing, cross the lower weaver over the upper one, and hold in place with the left thumb and forefinger.

*Insert the right forefinger between the crossed arms of the weavers with the length immediately to the right in the palm of the right hand and held in place by the pressure of the third and fourth fingers against the palm. This leaves the weaving to be done by the thumb and first two fingers of the right hand. With the index finger between the weavers, make the movement with the hand as described. This will cross the weavers and bring the bottom one to the top. Move the left hand

forward to hold the cross, insert the spoke and repeat between *

Do not twist weavers together between spokes as this locks the weave.

As the weavers change position with each cross, they must be regrasped in the new position. Tension is kept even on the weavers by the pressure of the right forefinger, and a tight weave is produced by pulling on the weavers with the right hand against a weave held stationary with the left.

Splice the weaver butt to tip. In case of a too-heavy butt, as in cattail rush or other bulky materials, trim the end to a compatible width and twist the old and new strands together to make the new weaver. The splice is less noticeable if made at a cross of weavers.

The Slip-top Knot. This procedure may be used with other mediums but it is especially adapted to cane as it requires no crushing in tightening and holds its shape and position well.

After fastening the end of the weaver strand (this will usually be in the base weave), carry the strand to the position decided upon for the first knot in an ascending angle to the left. Hold firmly with the pressure of the left thumb and wrap it around the spoke pair.

Bring the strand under the left spoke of the pair, up between the pair, over the wrappings and the ascending angle to make a vertical top weave over horizontal wrappings of the weaver strand around the spoke pair.

Pull down snugly, bring the strand up at the right of the spoke pair, and slip the end under this top weave from right to left where it is in position for the next ascending or descending passage to a new knot location.

Four-strand Round Braid. This braid is unexcelled for hangers and is used for such a purpose in several of the examples in this work. It is actually the lanyard procedure known to every Boy Scout, but it is adaptable to mediums other than craft strip and to other widths of strands.

A soft material will crush to make a close-packed cord, and experiments with widths should be made to determine the width needed for a particular size of cord. Such a cord is the one of underwater cattail shown in Plate 5, Fig. 4. The material was so soft that the four ¼-inch strands were compressed by braiding into a cord ⅛ inch in diameter.

Usually, the width of the strand will determine the diameter of the tightly braided cord, but any cord made of strands of stiff material wider than ⅛ inch must be made over a core, such as a soft string of the proper size, to hold it in shape and prevent flattening and creasing. The string or padding should be fastened where the braiding strands are fastened, often under the neck stay strand, and where they end. The string will also help stabilize the cord should it be necessary to splice the cord strands. It is recommended that the worker master the braid by practicing with two different colored strips of cloth. When the braiding is correctly done, it will result in four rows of diamonds, two of each color, and any variation in pattern will enable the worker to see a mistake immediately.

For clarity in describing the procedure, we will assume the strips of cloth are colored black and red.

Loop the strips around a nail driven near the edge of a shelf or a hook at a comfortable working height. This forms four strands, black above red, with the black strands in the right hand. The colors must remain in the same hand throughout the braiding.

Slip the inside black under the inside red and back into the right hand. This results in loose upper loops with the lower or inside strands crossed red over black. The weave is made with the upper strands alternately in an under-two and back-over-one routine around the inner crossed strands.

Bring the upper black strand under the inner black and inner red strand and back over the red strand into the right hand.

Bring the upper red strand under the inner red strand and the inner black strands and back over the black strand into the left hand.

This is the entire routine and, once understood, it is very easy. Attention must be given to the making of a tight braid as it is difficult to tighten. It will be noted that one outer strand is always slightly higher than the other which identifies the next weaver. This fact should be remembered when monotone braids are being made and there are no colors to act as guide.

The braid should be perfectly round and very firm, which is accomplished by holding the strands tight and the weave close as it is being made.

If the strand must be spliced, slip the new strand into the weave over the old strand for a couple of diamonds and proceed as if both strands were one until the new strand is fixed in place, when the old strand may be cut off. Trim the new strand close to the edge of a diamond when the braid has dried.

Bamboo. Bamboo canes are found frequently in temperate-zone plantings, growing in great clumps. Bamboo is of rapid growth and very decorative.

Only mature, straight canes should be cut. Remove the bunches of leaf-bearing stems at the joints to less than 1 inch from the canes, leaving a more careful cutting away until the canes have dried. Tie the tips in small bunches and hang to ensure a straight product. If necessary, the canes may be laid on a floor but they will require turning occasionally. As with all curing of vegetable products, a dry, airy, sunless place is necessary.

When the canes are thoroughly dry, which will require a matter of months, cut them into the lengths wanted, carefully removing the rest of the leaf branches at the joint and avoiding any scarring of the outer finish. Split the lengths into the desired width with a little leeway for sanding, and reduce to the necessary thickness by paring away the excess from the inside to secure a thin, hardsurfaced product suitable for spokes and, when pared sufficiently thin, for weavers also.

Cattail Rush. Cattail rush is one of the prime handcraft materials. It is plentiful, can be found almost everywhere in one or another of its forms, may usually be had for the taking, is readily harvested and cured, and stores well. It is most satisfactory when harvested and cured where it is to be used, for it is extremely brittle when dry so that sharp bends are disastrous. With careful handling it gives highly satisfactory results and is a joy to work with. As previously noted, it grows in quiet water and bogs and should be harvested when the leaves are at their best as the tips begin to brown in late summer. It is easiest to harvest as well as cure if, with a sharp knife, the entire sheath is cut under the water just above the fleshy root. It must be handled carefully as any sharp bend which creases the green leaf will almost certainly produce a break when the weaving begins. Tying a long smooth stick or a narrow strip of board the length of the leaves in with them when they are being transported is good insurance.

Any dark, airy place furnishes excellent curing quarters if the sheaths are spread out or, preferably, hung so they have a good circulation of air. A much better color results from curing the leaves in the dark, and air is essential for a good product. Hanging them with the tip down is the ideal way to cure the leaves. This is easily done by tying three or four sheaths at each end of a short cord to be hung over a stretched line, rafters, or some such support. Once thoroughly cured, the sheaths will hold indefinitely if they are protected from moisture. Curing can also be accomplished by spreading on a floor, in which case turning occasionally is necessary.

Being very dry, the leaves are extremely brittle and must be carefully handled and moistened before they are used. In the moistened state they are pliable and very strong. The leaf is quite porous and becomes waterlogged by prolonged soaking. One of the easiest and best ways to moisten is to separate the sheath and lay the chosen leaves out straight on a large, heavy cloth and hose down leaves and cloth a couple of times. Wrap the cloth in such a manner that the leaves may be taken out as needed without unwrapping the bundle. Allow the leaves to lie overnight to become thoroughly pliable. As only select leaves were wet, it makes no difference which one is taken out and this method allows the others to be kept moist until needed. If for any reason all moistened leaves cannot be used by the end of the second day's work, they should be dried to prevent molding and the cloth should be scalded before it is reused. Dried leaves may be remoistened. The cattail leaf is double-faced, filled between with air cells. Drawing the leaf from tip to butt between the folds of a tightly held damp cloth will cleanse the leaf and will also deflate the air cells, so that the leaf will coil more readily. This wiping down should be given as the leaves are taken out of the cloth just prior to use.

The underwater part of the leaf is always wide, as it wraps the base of the sheath and, to some degree, that part is controlled by the depth of the water in which it grows. It is of a different color and texture from the rest, so it must be removed for the sake of consistent color and size of coil. It should not, however, be discarded, as it is very strong and pliable and may be split for use in jacketing small bottles. It is a delightful material with which to work, pleasant to both eye and touch. It splits evenly so it may be used on smaller forms.

Cornshucks. Where corn is stored unhusked or in the shuck, this material will require only selection, as the corn and shuck have been allowed to dry thoroughly in the field. For length of product choose long ears. Discard the outer shucks which will be coarse and usually discolored. Select from the inner shuck for a bright even color and texture with as few flaws as possible.

Honeysuckle Vine. Honeysuckle vine is one of the prime favorites with many native craft basket makers. It grows in gardens in many localities, especially in the South. It is hardy, of rank growth, roots readily, and frequently escapes to the wild where, if the conditions are favorable and it remains unchecked, it takes over the area. Consequently, there is no lack of raw material.

Only mature runners of small size with no branches should be harvested. The current growth is tough so it does not break readily but it will not hold its form and flattens in the curing process. It

may be used in other procedures but will not serve for the Solomon's knot which requires a very stiff strand to hold the curve yet be manipulated with ease.

The vine may be harvested at almost any season of the year, though the late summer or the fall will provide more completely matured runners with fewer branches. The joints of these should be cut out, or, if not too extensive, trimmed down to the size of the runner. The aim is for a small, smooth, long strand.

If a rustic effect is wanted, the harvested vines should be coiled very loosely and hung in an airy place to dry with the bark left on.

For a finished product the coiled vines must be boiled in a covering bath to which a little concentrated lye has been added, to loosen the bark and to toughen the vines. Allow to cool in the bath overnight, then draw the vine, one piece at a time, through a coarse cloth held tightly around it. Scrape off all bark particles and rough places. Wash thoroughly and allow to stand in clear water overnight. Wash again, coil loosely, and dry in the shade. Do not hang on a nail for fear of rust. Once cleaned and dried, the vines may be held indefinitely and are ready to use when moistened.

Mulberry-root Fiber. The root of the common purple-fruited mulberry tree furnishes an almost unbreakable fiber even when stripped fine. It is therefore an excellent tie to hold the weave in place around the necks of bottles for a decorative finish, as it is not bulky. It also is useful where any pliable binding is needed, for it has a color which blends with most natural-colored products, is easy to control, and dampens readily. Long strands can be used as whipping strands to bind coils together or to make the weaver coils themselves.

The fiber is not a commercial product so the user has the privilege as well as the necessity of harvesting his own material.

The root has a worthless woody core which is very brittle. The thick white inner covering lies beneath a very thin bark of bright orange which breaks up easily. The best method of harvesting is to dig and sever a root 1 inch or more in diameter at the end near the tree and pull out several feet of it. Wash well and split the covering down each side of the root so the core can be removed.

Insert the point of a knife under the inside edge of the inner layer and pull the loosened piece of fiber toward the tip end. Narrow raffialike strands will split away readily and dry quickly when they are exposed to the air. Splitting is best done when the product is fresh. Once dried, it will hold indefinitely. When moistened for use, it can be subdivided further as needed. By stripping from the inside, white material only is harvested with none of the scaly outer particles, which are inclined to cling.

Palmetto Frond. The sabal, tree palmetto, or cabbage palmetto as it is familiarly known, is native to much of the Southeast and Gulf Coast states where it filled many needs in early days. There are still some who fashion braids for hats and other household uses from the frond but they are in the main tourists or a few others who find it furnishes a pleasant hobby characteristic of the area. It is recommended that the introduction to Chapter 3 be read in this connection.

The fan should be cut when the leaf stem can be seen above the crown of the plant and while the leaf is still folded tight. The length will vary with the plant, the location, the weather condition, and the time of year, although it will have a very satisfactory length of 3 feet in any case. Spring fans will usually be longer and heavier than those of the dry summer and fall months. As the light of a full moon added to the sunlight often causes the fan to open its folds before it fully emerges from the plant, suitable fans can be found more readily if this fact is taken into consideration. This palm is known as the tree which wears boots, as the old leaf stems which are called boots cling to the trunk over a period of years. These furnish a handy ladder, enabling the harvester to reach the fan, pull it outward and downward, and cut the stem at the point of bend where the fibers are taut. This is easily done even with a pocket knife.

The folds of the sharply creased fan should be torn apart on the shorter fold divisions, leaving the double fronds, the ones with the heavy midrib, intact. Hang the leaf stem up in a dark airy place for six weeks or more depending on the weather conditions which will influence the drying process. If the curing location is dry, airy and, preferably, dark the fans may be held indefinitely.

The scrub or saw palmetto and the silver palm have much smaller leaves with a different shape, in that they have no leaf midrib, but they should be harvested and cured in the same manner.

To strip the frond, cut a double frond from the dry leaf and insert a strong, slender needle through the frond by the side of the midrib and strip frond from midrib by pulling the needle toward the tip. This will produce two fronds, of which the section nearest the midrib is longer and stronger.

Divide the half fronds into strands of the width desired by inserting the needle and drawing it to the tip. The frond tapers and so will the strand but a slightly longer sustained width may be secured by stripping the butt or base end just a trifle wide, so as to have proper strand width fall a little higher up, and then removing the excess width at the base in a very narrow, tapered lower section. This should be started with the needle but pulled away gently, equalizing the strand. A little practice will make this clear and prove its worth. The extreme tip becomes very narrow, and in addition it is weak, so judgment should be exercised as to how much of it is used, though a too-narrow tip will usually determine this.

Palmetto will break if it is manipulated or creased when dry but it is surprisingly strong and pliable when it has been moistened by soaking it in cold water for about fifteen minutes and wrapping it in a wet towel for several hours or overnight. Long soaking waterlogs the frond so that even though it may be made up very tightly, the evaporation causes a shrinkage which opens up the work disastrously. If the frond is held in the wet towel for more than forty-eight hours in hot weather, it usually sours and becomes permanently discolored. It is better to moisten in small quantities as the work proceeds. If for any reason the frond cannot be used in this time, it may be dried and remoistened as usual.

Pine Needles. Pine trees are often allowed to remain on the grounds of newly built homes but it is seldom that the best length of needle grows so close at hand. As the length of needle varies with the individual tree, a search should be made for one which grows needles no shorter than 14 inches. Of course shorter lengths may be used but they require more frequent splicing. Locating the right tree is not as difficult as it probably sounds for, even in towns which are in the longleaf-pine area, sidewalks often serve as a harvesting floor where the product may be had for the effort. The needles cure on the tree and drop as Nature has the tree discard them. They should be gathered when rain and dew have evaporated, allowed to lie loose under shade until thoroughly dry, then bundled for storage. In this condition they will hold indefinitely.

Umbrella-plant Stems. The umbrella plant of our grandmothers' gardens is still sometimes found today and has a long stem which, when mature, needs only cutting and hanging in the shade to dry. It is handled most readily by tying the stems loosely in bunches around the leaves, which are later discarded. This allows the entire stem to hang and cure straight.

Virginia Creeper. The Virginia creeper, sometimes cultivated but usually claiming squatter's rights, is a free-growing vining plant with long runners. These should be harvested carefully to avoid injury. Coil loosely for cleaning and curing as directed for honeysuckle vine.

This plant is much like the vining plant commonly called poison oak but it may be differentiated instantly by the fact that it always has five leaflets to its compound leaf while the poison oak has only three.

Wiregrass. Wiregrass grows abundantly in the savannas of the Southeast coastal region. Along with the pine needle, with which it is frequently combined, it is in the first rank of basketry materials and adapts readily to bottle-jacketing procedures. It should be harvested carefully by pulling a few mature rods at a time from near the center of the great clump the plant makes after years of growth, at the center of previous growths which have matured and hang down around the plant. Pulling selectively from the center will assure a young but mature product. Tie in bundles about the center of the length using a soft string and hang in an airy, shady, dry place. When cured it should be a soft grey green, and when dried it holds indefinitely.

Yard and other grasses. These require only to be harvested when mature and not damp. As blades are usually thin, they dry quickly in the shade and air. Not all, but some varieties will hold their color, a highly prized attribute. It is well to remember that in curing vegetable products most varieties respond well to careful harvesting when mature and the protection of a dry, well-aired, shady location.

Washing Jacketed Forms. It may seem that forms which have been jacketed will present a cleansing problem but they may be washed as easily and safely as any other basektry product. In fact, most vegetable products benefit by a periodic bath which not only cleans but helps preserve, as it returns moisture to the fibers and retards brittleness. The washing should be done quickly on a warm bright day in a mild, soapy, warm bath and any scrubbing done gently with a very soft brush. Rinse thoroughly, blot the extra water from the form, and dry in the shade in an airy place. Turn or prop the bottle so that the air can reach all parts of the jacket, for if it is set on its base, that part will not dry quickly and will likely discolor. When the jacket is thoroughly dry, polish any exposed glass with facial tissues.

Cleaning and Waxing Gourds. Gourds have a thin, skinlike outer covering which protects them to some extent while they are growing and dries with them. Even so, it is rare that one does not have some blemish. Spraying was not generally practiced in the old days and certainly not on gourds. Sometimes there was a patch of mildew, and this is a permanent discoloration. Other blemishes may be an uneven shape caused by the way the gourd lay or what it rested on while it was growing, or the results of insect attack. Sometimes a piece of smooth wood was placed under the growing gourd to protect its base if it showed promise of being a fine specimen, but choosing the best available was the usual practice.

In order to avoid warping, gourds must be thoroughly cured and dried before they are cut and cleaned.

Soak the gourd for a couple of hours or more in cold water, turning it frequently, or drape a heavy wet cloth over it to pull it down into the water, for it will not sink.

When the thin outer film will slip, give it a good scrubbing with a coarse cloth. Avoid all abrasions, for these will damage the finish and will show through the waxing to be put on. Any rough spots or blemishes may be scraped gently with a paring knife held so that it does not scratch.

Cut a paper pattern of the scalloped opening desired and cut a hole in the center of the pattern to fit around any neck there may be on the gourd. Fit the pattern neatly on the gourd at the scallop line, being sure that the center of each scallop is the same distance from the table on which the gourd is sitting in order to get an even-topped bowl. Any inequality may be left on the top section where it can be trimmed off less obtrusively, should the top be finished for serving duty.

Fasten the pattern to the gourd at intervals with cellophane tape and trace around the edge with a pencil.

Follow this tracing with a razor blade and cut entirely through the gourd for a space long enough to admit the point of a fine keyhole-saw blade. If the hard outer surface was penetrated by the razor blade, following tracing will be successful and comparatively easy, but due care must be exercised especially at the sharp points between the scallops and at the end of the sawing, where there is danger of a break or a chip.

Lift the top section very carefully and remove all seed and membrane. Do not attempt any scraping on a dry surface. Leave in what does not come loose readily and fill the gourd with soapy water until the soft inner surface of the gourd can be scraped out with the edge of a tablespoon.

When all of this pulp has been removed, rinse thoroughly, fill with diluted chlorine bleach, and soak again until the inside is white. In order to bleach the points above the water line, moisten them frequently while the gourd is bleaching. If more bleaching is needed for these points, turn the gourd upside down in the chlorine solution which has been made just deep enough to cover the points. Oversoaking is not recommended for the sake of the outside finish.

After the bleaching process, rinse the gourd in several baths and immerse it in clear water for twenty-four hours.

Drain and allow to dry thoroughly in the shade, which may take a period of several weeks. Place so that the air reaches all surfaces, even the base.

When the gourd is dry, sand down any slight irregularities or blemishes with fine sandpaper and bevel the edges of the scallops slightly toward the inside. The inner surfaces should be sanded also, though a heavier grade of sandpaper is permissible here for the greater part of the sanding. Finish off with fine sandpaper for a smooth, hard-shell surface.

The waxing is baked in and applied to both outer and inner surfaces. This will require a good deal of attention but it does not have to be done all at one time as it can be worked in among other kitchen duties over a period of any length.

Heat the oven and lay a heavy brown paper on the open door. Provide a pound of paraffin and melt a small quantity of it. With a small mop, apply the melted paraffin to the side next to the heat, then turn and apply wax to another section, until the entire outside has been given a preliminary waxing.

Mop the inside with a generous coating of wax, turn the gourd on its side to face the heat, and turn it frequently, as the melted wax will collect in the gourd and should be distributed to coat all inner surfaces. Bake the wax in well, set the gourd up

right, and give the outside another application of wax, baking it in well. There will be wastage of wax but the brown paper will hold it and protect the oven door. Continue the applications of wax to the outside and the inside of the bowl alternately, baking it in well between applications until the entire shell is saturated.

When it dries, rub briskly with a soft cloth to give a polish.

The inner surface of the top shown as a server in Plate 1 had such a lovely texture under the sheets of membrane that these were removed very carefully and the inside left without scraping, since no liquid was to be served in it. In all other respects, however, the cleaning and bleaching was the same as for the bowl, and a good coating of wax was applied to the outer surface of the top and baked in lightly, although no effort was made to secure a deep waxing.